CW00408422
WORLD
BIG
IDEAS

SMALL WORLD BIG IDEAS

ECO-ACTIVISTS FOR CHANGE

Edited by Satish Kumar

First published in the UK in 2012 by
Leaping Hare, an imprint of The Quarto Group
This edition published by Aurum in 2022,
an imprint of The Quarto Group.
The Old Brewery, 6 Blundell Street,
London, N7 9BH, United Kingdom.
www.Quarto.com/Aurum

Published in association with The Resurgence Trust
www.resurgence.org

British Library Cataloguing-in-Publication Data
A catalogue record for this book is available from the
British Library

Jacket Design by Paul Smith
Typeset in Avenir and Bebas Neue by Claire Cater
Printed by CPI Group (UK) Ltd, Croydon, CR0 4YY

ISBN: 978-0-7112-7537-9
E-book ISBN: 978-0-7112-7538-6

“THE HALLMARK OF A GOOD ACTIVIST IS BEING ABLE TO EMBRACE THE DEEP HUMAN AND SPIRITUAL VALUES OF RESPECT, APPRECIATION, KINDNESS AND HUMILITY – WITHOUT THESE VALUES, AN ACTIVIST WILL NOT BE ABLE TO TOUCH THE HEARTS AND MINDS OF OTHERS”

SATISH KUMAR

CONTENTS

INTRODUCTION

by Satish Kumar

Every acorn is a potential oak. And if the right conditions of soil, water and sunshine are provided, something as small as an acorn will become a mighty oak tree. In a similar manner, every human being, however insignificant he or she may think himself or herself to be, is a potential activist – provided the right conditions are met to develop the qualities they will need of courage, commitment and selfless service. Just as an oak offers shade for the weary traveller, a branch for a bird's nest or a beam for the farm barn, an activist cares for the Earth, serves the poor, liberates the oppressed and achieves great heights of both imagination and self-realization.

A true activist is not a rare hero, nor an ego-driven dictator, nor a self-conscious superstar, nor a self-centred celebrity, nor a power-manic manager, but a humble host to humanity – a servant of the Earth and an ever-vigilant conscience of all the people. Such an activist is as mindful of the process and purpose of life as she or he is aware of the goals; there is no conflict between the means and the ends here, there is complete harmony between what is to be done and how it is to be done. True activism, therefore, is about big vision and right action rather than about outcome, achievements and unrealistic targets. A real activist lives by example. Anyone who demands 'Do as I say and not as I do!'

is not a good activist. Integrity between words and deeds is an essential quality of inspirational activism. Mahatma Gandhi was once asked. 'When you call upon people to do something, they follow you in their millions; what is the key to your success?' Gandhi reputedly replied: 'I have never asked anybody to do anything that I have not tried and tested in my own life. We have to practise what we preach. In other words, we have to be the change we wish to see in the world.' One living example is more effective than a million words; congruence between preaching and practice is a prerequisite for purposeful activism. We are all potential activists. We can show the world that a good life can be lived without exploitation, subjugation or domination of others, or of natural resources. We can show that a simple, wholesome and equitable life can be joyful and good.

We can show that happiness doesn't flow from material goods or the amount of money in our bank accounts: rather, happiness flows from the quality of the life we live, and the kind of relationships we have with our families, with our communities and with the natural world. This is bottom-up activism. We don't have to wait for a Messiah. The end of apartheid in South Africa, the establishment of civil rights in the US, the dismantling of the Berlin Wall, the disintegration of the former Soviet empire and many other such transformations occurred in the history of humanity solely because millions of people decided to take action at grassroots level – they refused to accept the unjust order of the day. Eco-activism is the greatest and most powerful example of people taking personal responsibility to participate in the process of the great transformation required for a just, sustainable, regenerative and resilient future for the Earth and her people. The environmental activism to address the climate emergency has become the biggest and all-encompassing human struggle of our time. Greta Thunberg, Extinction

Rebellion and many other leading scientists and activists, included in this book, have inspired hundreds of thousands of people of all ages and all over the world to come out in the street and be counted.

This radical and peaceful eco-activism has helped to create mass awareness among people at large, including politicians and business leaders, to develop new policies, new technologies and more importantly new attitudes towards Nature. Now more than ever increasing numbers of people are realizing that there is no separation between humans and Nature. Humans are Nature too. So what we do to Nature we do to ourselves. This rising new vision gives me great hope for the future. It is possible that humanity will rise to the challenge of climate catastrophe and collectively act to create a regenerative culture so that people and planet can live in harmony with each other and we all can look after our precious planet Earth for millions of years to come! In the Sanskrit language there is a word, *Sadhana*, which means a long, spiritual and practical practice and training through action and engagement. There is no shortcut to the development of qualities that lead to activism. It is a gradual process. An activist needs to be fearless, willing to suffer hardship and prepared to accept the consequences of adhering to his or her ideals. If you take the great examples of true activists like Mahatma Gandhi, Martin Luther King, Nelson Mandela, Aung San Suu Kyi and Mother Teresa, they all suffered hardship. Therefore the most important quality of a true activist is to live by one's ideals, whatever the consequences. Such activists are often criticized, ridiculed and ignored, but they do not give up. Qualities such as as resilience, endurance and integrity are the key to their successful activism. No activist can act by himself or herself; we all have to work with others. We have to recognize good qualities in others and avoid getting bogged down in, and preoccupied with, the negative behaviours or attitudes of other people.The hallmark of a good activist is being able to embrace the deep human and spiritual values of respect, appreciation, kindness and humility – without these values, an activist will not be able to touch the hearts and minds of others.

True activism is not about heroic headline-grabbing actions; it is about living and acting with integrity and without fear. Activism is, I believe, an inner calling. It is a journey. A journey of transformation from subjugation to liberation, from falsehood to truth, from control to participation and from greed to gratitude.

I have been privileged to know many great activists in my life. I was a student of Vinoba Bhave, who walked 100,000 miles for land reform; I met Bertrand Russell, who went to prison for peace, and Martin Luther King, whose dream was to bring an end to racial discrimination. I continue to work with similar, even if lesser known, activists who have dedicated and committed their lives to equally important and empowering ideals. And now I have the very great pleasure and honour of bringing together thirteen of these inspiring activists – some very well known, some less so – in this book. These men and women represent thousands of others around the world who are a 'salt of the earth' kind of inspiration to humanity and our great hope for the future.

This younger, Greta Thunberg generation, gives me great hope for the future. When they become the leaders and run political, social and commercial institutions they will run them differently. The economy under their leadership will be in harmony with the natural world. When this new generation takes charge, humanity will not see Nature merely as a resource for the economy, Nature will be recognized and respected as a source of life itself. Because of this hope I, even at age 85, am an optimist. To be an activist one has to be an optimist. You cannot be a good activist if you are a pessimist. I am still an activist, a happy activist. I urge all activists, young and old, to be happy activists rather than miserable activists! Act out of love, rather than out of fear or anger or anxiety; love of Nature, love of our precious planet Earth, love of humanity and love for future generations!

I offer their stories as a gift to you, the reader, so that you too may discover your own inner activist and, just like them, become a source of change and serve this small world we all share with your own big ideas.

SATISH KUMAR

Editor Emeritus of *Resurgence & Ecologist*, UK

Born in India in 1936, Satish Kumar is a peace activist, ecologist, Editor Emeritus of *Resurgence & Ecologist* (the flagship magazine of the Green movement), founder of the pioneering Schumacher College, and member of the advisory board of Our Future Planet, an online community sharing ideas for change.

A former Jain monk, Satish decided he could achieve more 'back in the world' – and has been quietly setting the agenda for global change ever since, starting with an 8,000-mile peace pilgrimage, recounted in *No Destination*. Satish Kumar edits *Resurgence & Ecologist* magazine from his home and its readership extends to twenty countries. Founded in 1966, *Resurgence*, as it was originally named, is thought to be the longest-running environmental magazine in Britain. The issues tackled when the magazine was first published were only just filtering through to the mainstream debate – endangered environment, renewable energy and ecological economics. *Resurgence* merged with *The Ecologist* in 2012, and continues to publish cutting-edge articles from a diverse range of contributors

In 1991 Satish Kumar co-founded Schumacher College, a 'green university' in south Devon. The college offers transformative learning for sustainable living, grounded in an ecological and holistic world view.

I am a peace activist. I come from a Jain family. Jains have made the meaning of peace as wide and as deep as possible. Jains were, and to a large extent still are, the original and staunchest pacifists of India; they have always advocated the practice of peace in such a manner that, from time to time, they have been labelled the practitioners of extreme non-violence. Jains are allowed to ordain monks and nuns at an early age. I became so deeply and profoundly attracted to our Jain guru, Acharya Tulsi, that I persuaded my mother to let me leave home to follow the way of the Jains and live and learn the principles of peace.

I was only nine years old and was fortunate that my mother – somewhat reluctantly – understood. She said to me: 'If that is your calling and your destiny, then who am I to be an obstacle to your spiritual seeking?' Other family members and some of my mother's friends were not so open or generous! 'How can a boy of nine know what is his calling or his destiny?' This was the argument my brothers made and the answer my mother gave them was filled with emotion and conviction: 'I know, I know. It is hard for me to let this little boy leave me, but a child is not an under-developed adult. If we dampen or discourage his desire to seek a spiritual life now, how do we know what will be the effect on his tender soul? It is not easy but on balance, we have to let him do what he wants to do.'

My brothers were amazed but my mother's words gladdened my heart. She loved me but did not want to possess me. I believe it was she who laid the foundation of courage and activism in my life by being bold and selfless enough to let me leave home and follow the path of peace.

I remained a Jain monk for nine years. It was a time of training in peace. The most important training of all was walking barefoot. For nine years, I rode no animals, travelled in no cars and no trains, no boats, no bicycles and no aeroplanes; and yet I walked thousands of miles – in hot and cold conditions, through deserts and jungles, across mountains and plains. I had no home and no money; no storage, no possessions. I begged for food once a day and ate what I was given, keeping no leftovers overnight. I possessed only what I could carry on my body. Walking was not just a means of travelling from A to B, it was not just a means of arriving somewhere. Walking was a spiritual practice in itself – a practice in simplicity, minimalism and meditation.

Monkhood, then, was my preparation for the realization of peace within. 'How can you make peace in the world if you are not at peace within yourself?' said my guru. 'Peace is not only what you think or what you say but how you are. Making peace begins with being peace; there is no distinction between the liquidity of milk, the whiteness of milk and the nutritious qualities of milk. The consistency, colour and quality of milk are an integrated whole; similarly, thinking, speaking and being peace must be integrated. Then peace will radiate from you as effortlessly as light radiates from the sun.'

When, at the age of eighteen, I left the life of a mendicant and joined another peace activist, a Gandhian man called Vinoba Bhave, I added a social dimension of peace to the spiritual dimension I had acquired as a young Jain monk. 'There can be no peace in the world while the powerful dominate the dispossessed and the rich exploit the poor,' explained Vinoba. 'India is now free from the British Raj and colonial rule but India is still not free. As long as landlords live in luxury and landless labourers toil for starvation wages, there can be no peace even if there is an appearance of calm.'

With this conviction, Vinoba was going from landlord to landlord asking them not to wait for an armed uprising or government legislation, but to act out of the compassion of their hearts and in response to a political call for social justice. He said they should act right away to make peace with their neighbours and peace with their peasants by sharing their land with them. This was a movement of Land Gift to establish political peace.

I was so deeply inspired when I met Vinoba that I decided, as I had no land to give to the cause, to give instead the gift of my life. Vinoba, like me, was walking. Jeeps and cars cannot reach where your legs can reach – not only the remote villages and hamlets, but also the hearts of people. I was ready for that. I had already walked thousands of miles so my feet were toughened, my muscles were hardened, and my resolve was resilient.

'What should I say to the landlords to persuade them to give up their most precious possession – their land?' I asked Vinoba. 'Tell them, if you have five children, consider Vinoba your sixth child, representing the poor, the wretched and the weak. Give one sixth of your land. If one sixth of the land of India was redistributed among the landless farm labourers, then no one would be left without a source of livelihood.'

'But this seems too idealistic, Vinoba,' I protested. 'People are bound to resist, thinking, "We own this land, we inherited it from our parents, it is our land, why should we give it away?"'

Vinoba replied: 'How can anyone own the land? It belongs to Nature. Do you own the air? Does anyone own water? Does sunshine belong to anyone? If air, fire and water cannot be owned, then how can the Earth be owned by anyone? We cannot claim ownership of the basic elements of life, we can only live in relationship with them.'

Vinoba continued: 'In any case, if we do not share the land then the poor and the dispossessed are not going to take it lying down for ever. Do we want a violent revolution? Is it not better to bring about change in a sensible, rational and peaceful manner?'

Equipped with such sincere, uplifting and convincing thoughts, I went, with fellow peace activists and campaigners, to a landlord who was at the same time the head of a Hindu Temple owning 2,000 acres of prime paddy fields.

But how would we get to see him?

'Please tell the High Priest that some messengers from Vinoba wish to see him,' I told the staff at the Temple office.

The staff frowned but offered us a glass of water. They knew that Vinoba was no ordinary person; he was, apart from being a radical land reformer and a peacemaker, also a renowned Hindu scholar whose talks on the Gita, one of the sacred texts of India, have sold millions of copies in more than a dozen languages – and yet we were refused an audience with the Temple's High Priest.

It was not that he did not have time to see us, rather that he did not wish to entertain the idea of parting with some land in favour of the landless. And so even before we had the chance to present the vision, the values and the ideals of land reform through peaceful means, we had been rejected outright.

What then should we do?

The following day we prepared some banners and a dozen of us returned to the Temple office very early in the morning and before the High Priest arrived. Our banner had no radical slogans – the only demand we printed on it was: 'We wish to meet the High Priest.'

As we saw the Jeep that he travelled in approaching the Temple, we moved to block the gate. And because the Jeep had no doors, there was no physical barrier between us and the High Priest. We surrounded the vehicle so it could not move forwards or backwards.

'What do you want?' the High Priest asked. 'We want an audience with you.'

'But I am very busy,' he replied.

'But the poor are very hungry,' we said.

'I will arrange to provide food for them. Now, let me go.'

'But they will be hungry again tomorrow, and the day after, and the day after that. How long can you go on feeding them? We want you to do something that will enable them to feed themselves.'

The High Priest fell silent.

'I know what you want. You want me to give land,' he said.

'Yes, that's right. Then the poor and the hungry can feed themselves for ever and without disturbing or bothering you. You will gain peace of mind and you will be living in a happy neighbourhood. For yourself, you will gain merit and a good reputation. People will sing your praises. When you are kind to the poor and generous to those who have nothing, you yourself will be happy.'

This is how we pleaded with him.

'I have to consult with my committee. And in any case, why don't you yourselves raise some funds to buy the land? I would certainly contribute towards such a fund. I am sure we could give 1,000 rupees; even 5,000 rupees. That would be a much better way. I just don't think we can give you land,' the High Priest told us.

We were delighted that we were having such a dialogue with him at the gate to the office near the Temple. People from the street had started to gather around to listen and the Temple officials were looking worried.

'Please clear the way, the High Priest must attend to his appointments,' said one of those officials. But we paid no heed to his request.

I said to the High Priest: 'There are millions of people in our country who have no home, no land and no livelihood. How much money can we raise to buy the land for these teeming millions? This is why Vinoba, in following a spiritual solution, is advocating Land Gift. He says money is the problem, not the solution. Gift is the solution. The giving of a gift is as beneficial to the giver as it is to the receiver.'

'Let me think about it,' the High Priest replied.

'Can we come and see you tomorrow morning? That will give you twenty-four hours to think,' I replied.

'We are not asking for anything for ourselves. Vinoba has no home, he lives in an ashram and walks around the country collecting land for the poor. So please remember, giving the land to Vinoba for the poor is as much in your interest as it is in the interest of the recipients – the landless labourers and farm workers. 'Please also remember Vinoba's message that the land does not belong to us, we belong to the land. When we die, we do not take it with us. This is why Vinoba is asking us to share and care – so that you and your workers are happy together.'

I had found myself in a strange position, reminding the High Priest of a Hindu Temple of the highest of Hindu ideals.

He smiled at me and said: 'Alright, come and see me tomorrow morning. We will see what we can do.'

We clapped. Everybody clapped. Yesterday, we had wanted a private audience with the High Priest. Today, we had been given a very public audience, which had proved much better for us.

The next morning, the staff at the Temple office were much more cooperative. As soon as we arrived, we were given glasses of water and cups of tea; and very soon we were ushered into the presence of the High Priest.

Before we could utter a single word, the High Priest told us: 'I know Vinoba is a great spirit, he is a man of compassion and commitment to the uplifting of the poor, so please take this note from us to Vinoba.'

The note said: 'We wish to donate 120 acres of our land to Vinoba for redistribution among the landless farm workers. Please arrange for your colleagues to work with our staff to locate the area of land precisely and mark it out.'

It was an easy success for us and we were over the moon.

This was now happening all over India. Between 1951 and 1971, four million acres of land were donated to Vinoba in a similar manner. This was social activism at its best. And during those years, Vinoba himself walked over 100,000 miles covering the length and breadth of India; from Kashmir to Kerala, from Goa to Assam. And I had the great good fortune of walking with him and working with him as an apprentice activist.

One morning, in 1961, my friend E.P. Menon and I went to a café for our morning coffee and breakfast. At this time, we had both been assigned by Vinoba to work at an ashram near Bangalore, a city of culture and commerce with a cool climate and famous cafés and restaurants. We were deeply grateful to Vinoba for giving us this opportunity to establish an ashram where young activists, like ourselves, could be trained and inspired.

On this particular day, we were to encounter a piece of news that would change the course of both of our lives and change them forever. While we waited for our coffee, we read in the newspaper that a ninety-year-old philosopher and Nobel prize-winning mathematician had been sentenced to a week in jail for protesting against nuclear weapons in Britain. Reading such an extraordinary story, we were surprised, amazed and shocked – all at the same time. Menon and I said to ourselves: 'Here is a man of ninety going to jail for peace in the world. What are we, as young men, doing sitting here drinking coffee?'

The man was, of course, the British peace activist, Bertrand Russell. We felt challenged to do something to support this philosopher who had shown such enormous courage and conviction – but do what?

We then learned that Bertrand Russell and thousands of other protestors against nuclear weapons were marching from Aldermaston to London and so we thought, why don't we march too? Not only to London, but also to Moscow, Paris and Washington – in other words, to the capital cities of all those countries in possession of nuclear weapons at that time.

'Wow!' We both surprised ourselves with the idea. We'd found our inspiring idea – we were literally going to walk to all four nuclear capitals of the world.

We flew to Assam, where Vinoba was walking for Land Gift. It was not hard to find him – both the newspapers and the radio covered his every move and anyone who was anyone knew where he was. He was not only the talk of the town but also the talk of the state of Assam!

We reached Vinoba in a remote village where he had just arrived after walking for 10 miles. When he saw us he said: 'What brings you so unexpectedly? Are you ok? How is everything and everyone in Bangalore?'

We were apprehensive. After giving news of the ashram, we revealed we had come specifically to ask for his permission, support and blessing to undertake a mission for world peace and the cessation of nuclear weapons. Vinoba was a true saint – selfless, detached and unperturbed.

'Interesting!' he said. 'Will you walk all the way? Which route will you take?' He asked us to show him a map of the countries through which we would walk.

We held a long conversation about the route and the advantages/disadvantages of various alternatives. Then, suddenly, he pushed the map aside.

'Yes, you have my blessing, but I give you two weapons to protect you on this long journey. Firstly, you are vegetarians and you must remain so, and secondly, you must walk without a penny in your pockets.'

We were stunned.

'Remaining vegetarians we understand, Vinoba, but when you say take no money, do you really mean we must have no money at all? What if we need to buy a cup of tea or make a telephone call?'

Vinoba told us: 'Either you should have a lot of money, like a king, or you should have no money and so live like a *sadhu*!'

He laughed and continued: 'War begins in fear of others and peace emerges out of trust. It is no good you just preaching peace – to be a true peacemaker you have to practise peace, which means practising trust. So, you go on this great journey with trust in your hearts. Trust people and trust the process of the universe and all shall be well.'

But we were still apprehensive and unsure.

'Surely, we trust people but for convenience sake, could we not carry just a little money with us?'

Vinoba gave the practical reason for walking with no money at all. 'When you arrive in a place after a long walk, you will feel tired and exhausted.

You will eat in a restaurant, sleep in a guesthouse and walk away the next day; but when you have no money, you will be forced to find someone who can give you hospitality for the night – they will offer you food and you will say you are vegetarians. And they will ask why? And then your communication about peace can begin.'

Vinoba was my guru, and the relationship between a guru and a novice is of complete and unquestioning surrender to the guru's guidance and wisdom. In any case, having lived without money for nine years as a Jain monk, I was marginally more prepared than Menon – but even he was courageous enough to say: 'Yes, let our journey be a pilgrimage.'

Having received the blessings of Vinoba, we went to New Dehli, to Raj Ghat, to the grave of Mahatma Gandhi, and began our pilgrimage not only without money but even without passports. The Indian government in those days required us to place a deposit of 20,000 rupees in case we had to be repatriated for any reason; the deposit would be there to cover those costs. This was a good test for us – whether or not to trust the Indian government. The Indian newspapers covered the story and a member of parliament raised the question with the Prime Minister, Mr Nehru, himself: 'Why are peace pilgrims denied their birthright of free citizenship, unrestricted travel and a passport?'

Mr Nehru at that time was not only the Prime Minister but also the Minister of Foreign Affairs. We had already informed him of our plans and we had received an encouraging response from him, and also his good wishes. So he took personal responsibility to cut through the bureaucratic red tape and on the day before we were scheduled to cross the border into Pakistan, two officials arrived in search of us and delivered our passports. Trust had prevailed.

Thirty-five men and women came to bid us farewell; most were enthusiastic about our plans but one woman friend of ours was rather worried and concerned. 'Aren't you crazy, my friends, to walk into Pakistan? We are at war with that country, they are our enemies. You have no money, no food and you are on foot. What about your safety and security?'

She spoke to us scoldingly: 'Forget Vinoba, at least you should carry some food with you. Here are some packets of food to sustain you while you are looking around for a sympathetic host.'

This was a moment of trial. I thought for a minute, reflecting on Vinoba's words. I said to my friend: 'I thank you for your kindness but these packets of food are not packets of food, they are packets of mistrust. What are we going to say to our Pakistani hosts? We didn't know whether you would feed us or not so we have brought our own food all the way from India. So please understand and forgive us for refusing your kind offer.'

The woman was in tears.

'Why are you crying, my friend? Please give us your blessing.'

'Satish, this might be my last chance of seeing you; you are going to Muslim countries, Christian countries, communist countries, capitalist countries, deserts, jungles, mountains, snowstorms and you have no food and no money to support you. I cannot imagine that you will survive this ordeal and return alive.'

My friend was really sobbing now.

'Don't worry, my dear friend. If I die while walking for peace, that is the best kind of death I can have – so from today, I will walk for peace without fearing death and without fearing hunger. If I don't get food some days, I will consider it my opportunity to fast and if I don't get shelter some days, I will take it as an opportunity to sleep in a hotel of a million stars. Surely that will be even better than a five-star hotel.'

But even that joke did not satisfy my friend. She hugged me tight, still sobbing, but she also realized there was no way to stop me. Life or death, Menon and I were determined to go.

To our great surprise, as soon as we stepped onto the soil of Pakistan someone approached us and enthusiastically asked: 'Are you the two walkers coming to Pakistan for peace?'

'Yes, we are, but how did you know? We don't know anybody in Pakistan. We've written to nobody in your country and yet you seem to know all about us. How come?'

'Your fame has travelled ahead of you. I read about you in the local papers and some other travellers, also coming from India, had seen you and talked about you. When I heard that there were two Indians putting out their hands in friendship, I was touched and moved. I have been looking for you for days. I am also for peace. What nonsense that India and Pakistan should be at war.'

The stranger continued: 'Before 1947, we were one people. We cannot choose anything other than to be neighbours and friends and so I have come to welcome you.'

His words were music to our ears. We told him: 'Thank you, you have a big heart and a big mind. Peace cannot be achieved through mean-mindedness. If we come here as Indians, we meet Pakistanis. If we come as Hindus, we meet Muslims – but if we come as human beings, then we meet human beings.'

We carried on explaining. 'Being a Hindu, a Muslim, an Indian or a Pakistani, these are our secondary identities – being members of the human community and the Earth community is our primary identity.'

Our stranger host gave us both a hearty hug and offered hospitality at his home. Five minutes earlier, my friend had been sobbing, filled with fear of Muslim Pakistani enemies and here we were, just moments later, embracing one!

This was repeated day after day and night after night across Pakistan. Again and again, people proclaimed that war is not between Hindus and Muslims or between Indians and Pakistanis – war is between power-seeking politicians and profit-seeking manufacturers of the weapons of war.

Whether we were walking on the plains of Pakistan or the heights of the Khyber Pass, whether we were being entertained by the Rotarians of Rawalpindi or by the spirited Pathans of Pakistan, the cry of the ordinary people in the street was the same – don't waste our wealth or talent on weapons of hatred, do something to enhance the harmony and well-being of those who work in the fields, the factories, the schools and the hospitals.

Blisters aside, we were blessed by the generosity of human hearts throughout our journey, crossing the high hills of Afghanistan, the sandstorms of Iran and the lush vineyards of Azerbaijan. And having no money proved to be a blessing, rather than an impediment. The moment people realized that we were pilgrims of peace and had renounced any dependence on money, they were more eager than ever to help us and much more eager than if we had money!

The cool stillness of the Caspian Sea, the imposing grandeur of the mountains of Ararat, the sumptuous orchards of Armenia, the deep-green tea gardens of Georgia were as inspiring as the people of those countries. I have written about the transformative experience of the kindness we received during our journey in my book, *No Destination*, but I would like to retell the story of the 'peace tea', which gave the whole pilgrimage a particular focus.

Black Sea on the left, Caucasus Mountains on our right, we trudged along, day after day. 'Are we really achieving anything?' I asked my companion Menon, with no little desperation in my despondent voice. 'Are you feeling low?' he asked. 'Remember, it is all about action and not about results. Come on, pull yourself together.' As Menon was speaking, I was reminded of the famous saying of Rabindranath Tagore.

> 'If no one comes to your call, even then, walk
> alone, walk alone even if everyone looks away
> speak alone, speak alone even when the path
> gets tough tramp alone, tramp alone…'

Tagore's words helped me somewhat but not enough. I was still overcome with doubt. However, just then I noticed two young women standing, enjoying the sunshine. I gave them a leaflet, written in Russian, which described the purpose, the route and our action for peace.

One of them said: 'We heard you on the radio, what a coincidence that we should meet you. Have you really walked all the way from India?'

'Yes, we have!'

'Our saint Rasputin went to India on foot. Are you making a return journey?' 'You could say so,' we replied, in the basic Russian we'd been learning.

We were about to walk away when one of the women said: 'We work in a tea factory. It is our lunch break, would you like to have a cup of tea in our canteen and tell us all about your journey?'

'Of course,' we replied. 'Any time is tea time!'

We followed the women back to their factory, dropped our rucksacks to the ground and started to relax over a cup of tea. Some of the Soviet workers started to gather around us – seeing strangers from India was something very novel to them. Soon biscuits and bread also arrived and Menon and I were delighted to sit down and enjoy the delicious cup of tea, answering question after question.

The curiosity of these workers was boundless. This may have been a remote, rural location but their concern for peace and their amazement at the stupidity of the militarists wasting resources on useless nuclear weapons filled the room.

While we were deeply engaged in a discussion about disarmament, one of the two women we had met originally had a brainwave. She suddenly stood up, went out of the room and returned, moments later, with four small packets of tea.

'I have a special request,' she said. 'These four packets of tea are for the four leaders of the nuclear countries of the world. I cannot reach these leaders but I want you two to be our ambassadors and please be the messengers of this "peace tea" and deliver one packet to our Premier in the Kremlin, the second packet to the President of France in the Élysée Palace, the third packet to the Prime Minister of the UK at 10, Downing Street and the fourth packet to the President of the United States in the White House.'

We listened to her request in complete silence. What an imaginative present. 'And please deliver the tea to the leaders with a message from us, from this little tea factory by the Black Sea. Our message to them is this: this is no ordinary tea, this is peace tea and if you ever get the mad thought of pressing the nuclear button, please hold for a moment and have a fresh cup of peace tea. This will give you a moment to reflect that your nuclear weapons will not only kill your enemies, they will kill all men, women, children, animals, forests, birds, lakes – in short, all of life. So, think again and don't press the button.'

'Wow!' I said. 'What a message.'

I told Menon, all my despondency had disappeared; come hail or high water, we would deliver these packets of peace tea to their desired destinations – just as the women had requested. And we thanked her for giving us such a special assignment.

'We are honoured to be your ambassadors of peace,' we told her.

The woman's face lit up. She was full of charm, grace and beauty. As she hugged us, everybody in the room clapped; this was the best send-off we had had.

With many ups and downs, struggles and strife, we finally arrived in Moscow.

The Premier, Nikita Khrushchev, sent us a warm letter congratulating us but regretting that he could not meet us personally. However, we were invited to the Kremlin to deliver the packet of peace tea to the Chairman of the Supreme Soviet, Mr Tikhonov, who would pass the gift to the Premier.

The splendour of the Kremlin was impressive, but what Mr Tikhonov said on receiving the tea was less convincing: 'Our Premier Khrushchev and the government are making proposal after proposal to the Western powers to safeguard peace.

So your work is really there, in Western Europe and America, and I am glad you are taking your message to them.'

Thus, he had passed the buck!

From Moscow, through deep snow, we walked. Through the Russian towns and villages as well as Belarusian, Polish, German and Belgian rural and urban landscapes. It had taken us ten months to get to Moscow and it took us another six months to reach Paris. We were eager to see President de Gaulle but all of our letters and telephone calls brought no response. So, with the support of the French peace workers, Menon and I went to the Élysée Palace seeking an audience with the President or his representatives. But the guards, the officials and the police all ordered us to move on as it was illegal to demonstrate or even assemble at the gates of the palace. As we refused to obey their orders, we were arrested and taken to prison, where we were threatened with deportation back to India.

Eventually, after three days of negotiations that involved the Indian ambassador, we were allowed to deliver the peace tea to the head of the Paris police, who promised to pass it on to the palace – and in a way, we were pleased to have been imprisoned in Paris since this really was our action, following in the footsteps of Bertrand Russell.

Helped by our French friends, we crossed the channel by boat and then walked from Dover to London, seeking a meeting with the Prime Minister, Harold Wilson. Again, the Prime Minister was too busy to meet with us; however, he asked Lord Attlee, the former Prime Minister, to receive us and accept the packet of peace tea on behalf of Mr Wilson. On receiving us at the House of Lords, Lord Attlee said: 'Dear chaps, be assured that no one is going to use nuclear weapons, it is only a show.' He laughed, but the words of this short, slim and serene politician did not convince us.

'If that is the case, then why are we wasting our time and resources on these toys while people in the world starve, schools and hospitals lack resources and children grow up in fear?'

'It is politics, youngsters, it's politics!' he replied.

It was all very well for a retired politician to be so complacent but he did promise that the packet of peace tea would have a place in 10, Downing Street, the home of the Prime Minister.

We met Bertrand Russell. That was a moment of exhilaration and inspiration. After our long chat and stories of our adventures, Lord Russell served us tea and cake, saying: 'Make tea, not war!'

We crossed the Atlantic aboard the *Queen Mary* liner and arrived in New York, walking from there to Washington D.C. President Johnson was in the White House; he appointed his Chief of Disarmament to receive us and he gave us a cordial welcome.

In comparison with the Kremlin, the White House felt new and far less spectacular; but the politics of the two places was not dissimilar.

'America leads the negotiations for peace at all international forums, it's the Soviets who create obstacles. As you have been there, I'm sure you know what they are like. Communism and peace are a contradiction in terms... what can we do?'

If President Johnson had a negotiator for disarmament with such a closed mind, what hope was there for peace, I wondered. Nevertheless, we were given an undertaking that the packet of peace tea would be cherished by the President.

We did our best to be the ambassadors of the tea producer by the Black Sea. And that potent peace tea is still working. No nuclear weapons have been used. And fingers crossed they never will be. However, these bombs are still there and as long as they exist, I will continue to act for their disarmament. That is my promise of peace activism.

And having walked 8,000 miles across fifteen countries, I realized that peace is a state of mind and a way of life. I started with a search for inner peace as a Jain monk, then pursued political peace with the Land Gift movement, and followed that with a search for world peace by challenging the nuclear powers to declare unilateral disarmament. During this time, I became aware that I am held by the Earth, nourished by Nature and sustained by rivers, forests, flowers and wilderness.

Unless we make peace with the planet, inner peace and world peace will remain elusive.

By walking across the continents and experiencing the workings of our modern civilization, I came to the conclusion that humanity is at war with Nature. The way we are extracting and using natural resources: cutting down the forests, overfishing the oceans, poisoning the land with chemicals, pesticides and herbicides, and emitting greenhouse gases into the atmosphere, changing the very climate that sustains our lives – all these acts are acts of aggression against the Earth.

Therefore, I am not only a peace activist but also an eco-activist, not only a peace pilgrim but also an Earth pilgrim.

And thus, my activism is inclusive of inner peace, world peace and green peace!

Further Reading & Useful Websites

No Destination: Autobiography of a Pilgrim by Satish Kumar (2014)

You Are Therefore I Am by Satish Kumar (2002)

Earth Pilgrim by Satish Kumar (2009)

Moved by Love: The Memoirs of Vinoba Bhave by Vinoba Bhave et al. (1999)

Talks on the Gita by Vinoba Bhave and Kamlesh D. Patel (2019)

www.resurgence.org
www.theecologist.org
www.schumachercollege.org.uk

FRANNY ARMSTRONG

Climate Change Campaigner & Filmmaker, UK

Born in London in 1972, Franny Armstrong is a self-taught and fearless filmmaker. Through her company, Spanner Films, Franny has produced the highly acclaimed documentaries *McLibel*, *Drowned Out* and *The Age of Stupid*, and also pioneered the 'crowd-funding' finance model, which enables independent filmmakers to raise workable budgets. In 1996, she founded the McSpotlight website, described by *Wired* magazine as 'the blueprint for all activist websites', and in 2009 she founded the UK 10:10 climate campaign, which launched internationally in 2011.

In July 2021 Franny Armstrong produced and directed *Rivercide*, presented by George Monbiot. *Rivercide* is a documentary investigation into the sudden rise in the pollution of Britain's rivers.

Born in the 70s, I'm part of the MTV generation who were told by a squillion adverts that the point of our existence was to play loads of computer games and go shopping lots. Finding out that this is most definitely not true – that the very future of life on Earth is in our hands – has been both overwhelmingly daunting and a bit of a relief. How so? Because every generation who came before us didn't know about climate change and everyone who comes after will be powerless to stop it. Our action or inaction in the next decade or so will not only define our generation in history but will, much more crucially, define whether humans are able to continue living on this planet – or not. Other generations came together to overturn slavery or end apartheid or win the vote for women or even land on the moon, so we know that immovable mountains can occasionally be moved. Our generation is not intrinsically more stupid or useless than our forebears, and there's no doubt about what we have to do. The good news is that we have all the knowledge and technology we need to avert disaster and the only thing currently stopping us is ourselves. But we have so far collectively achieved less than nothing – since the scale of the problem was first identified forty-odd years ago, total global emissions have always gone up and up.

I first heard about climate change as a teenager in a school biology lesson, when it was called The Greenhouse Effect. The teacher explained that we are conducting an experiment on ourselves by changing the life-sustaining systems under which our species and all others evolved. Nobody else in the class seemed too struck by what seemed to me to be an existence-realigning revelation. My immediate reaction was to launch a campaign that threatened to reveal which teachers were driving to school rather than biking or walking, as well as listing which of their cars were not fitted with catalytic converters (a simple, now-superseded technology that reduced the toxicity of the vehicle's exhaust). With typical teenage torpor, I never followed through with my threats, but years later bumped into a former teacher who told me that I had sent genuine fear through the staff room and that she and several others had converted their cars. It would be a long and winding twenty years before I worked out a more effective way to contribute to the battle to prevent climate catastrophe.

Then, shit happened. Cow shit, to be precise. My childhood ambition had always been to become a farmer and eventually I persuaded my mum to take us on a family holiday to a dairy farm. My sister and I astonished the rest of the family by leaping out of bed at 5 o'clock every morning to spend all day helping with milking, feeding and endless manure shovelling. But then, towards the end of the week, our favourite cow, Piggy – the solitary brown cow in a herd of Friesians – slipped on some shit and cut her own udder with her hoof. The farmer explained that it wasn't economical to pay the vet's fee of £37 to sew up the 2-inch cut and that it therefore made sense for Piggy to become beef straight away. As she was led away to slaughter, my sister and I became vegetarians on the spot, and have been ever since.

My decision to stop eating animals led to endless discussions at school and to more than half our class becoming veggie – twenty years later, I now know, thanks to Facebook, that three of them stuck with it – but I soon despaired of these circular conversations. As well as becoming bored with the sound of my own voice saying the same things, I'd get frustrated when I didn't convey my arguments perfectly or when someone brought in a new point for which I didn't immediately have a counter-argument. Light-bulb moment: changing the world one person at a time, via face-to-face interactions, ain't never going to work.

Meanwhile, my dad was a filmmaker making documentaries for the BBC about global injustices. I'd pretty much ignored his job up to this point – my brother, sister and I would sleep on the sofa through his programmes and our mum would wake us up when the credits came on so we could see his name go by. Then, for a once-in-a-childhood adventure, he took me with him to New York on a mission to sell his new film, *Global Report*, to US distributors. I sat at the back of the screening rooms watching the three-hour film over and over again, only occasionally making rabbit ears in the light of the projector. The film featured a young boy living in Tanzania who supported his family by selling peanuts in the streets, earning less than my weekly pocket money of 60p. 'So why can't all the English kids just send their pocket money to the African kids, Dad? That would solve the problem, wouldn't it?' This was my introduction to the concept of global inequality, hammered home by the opulence of America, which I was also witnessing for the first time.

Despite becoming increasingly politicized – volunteering for Greenpeace, helping with local recycling schemes, running the school's visit a pensioner scheme – it never occurred to me to follow in my Dad's footsteps and use film to catalyse change.

Still obsessed with animals, I studied zoology at college, and then volunteered for a six-month coral reef mapping project in Tanzania. While out there, my Dad told me on the phone about a court case that was just starting in London in which two local campaigners were standing up to the mighty McDonald's, who were suing them for libel after they distributed a pamphlet entitled 'What's wrong with McDonald's: Everything they don't want you to know'. 'It's very up your street subject-wise,' he said. 'Why don't you borrow my camera and make a film about it?'

Meeting up with the two 'McLibel' defendants back in London, they explained that eight documentary production companies were fighting over the rights to their story, including some of the biggest names in British TV. Clearly a first-time haven't-got-a-clue filmmaker like me had no chance. But they called a few weeks later saying that all those companies had dropped out after failing to get a commission to make the programme – the main TV channels had all had legal run-ins with McDonald's in the past and, funnily enough, were not falling over themselves to criticize Big Man again. 'But you have your dad's camera equipment, don't you?' said Dave Morris, one of the defendants, in a line that led to my becoming not only a filmmaker but an independent filmmaker, 'Why does not having any money mean you can't make a film?'

The next two years were a thrill-a-minute rollercoaster as my tiny team grappled with the legal, technical, logistical and moral complexities of turning a court battle involving spies, corrupt policemen, turncoat clowns and 60,000 pages of legal documents into a gripping ninety minutes of prime time screen time – all with no experience of making a film and funded entirely on fresh air and favours. It turns out that the phrase 'fire in the belly' is based on an actual physical feeling in the guts when you dedicate your waking – and often sleeping – life to a world-striding zeitgeist-sparking mission. By the time my filmmaking hero Ken Loach had come on board to direct the reconstructions of the courtroom, my head was spinning at the sheer wonder of being part of such an epochal story.

The resulting film, *McLibel*, was picked up by BBC1 and scheduled to be broadcast a few weeks after the trial concluded in June 1997. But, devastatingly for us, the BBC got cold feet and pulled out at the eleventh hour. Channel 4 then pitched in, but their lawyer said no. It was only eight years later, after the 'McLibel Two' had taken the British government to the European Court of Human Rights – and won – that our film was finally broadcast on BBC2, to excellent viewing figures (1 million at 10.30 pm on Sunday) and fantastic reviews: the *Guardian* called it 'absolutely unmissable' and the *Seattle Times* said it was 'an irresistible David and Goliath tale'. Upwards of 25 million people have now seen the film as it went on to be released in cinemas in Britain, the US and Australia, was broadcast on TV in twenty-odd countries and released on DVD worldwide. The biggest thrill came when the British Film Institute included it in their series 'Ten Documentaries That Shook the World' alongside classics like *Bowling for Columbine* and The *Thin Blue Line*. The only other British film chosen was Michael Buerk's original news report from Ethiopia, which led to Live Aid.

When I set out to make *McLibel*, I never for the slightest second thought we would have any noticeable impact on the corporate behemoth. I just found the story of two people daring to stand up to McDonald's enormously inspiring – and felt that others would too. But ten years later – thanks also to *Fast Food Nation*, *Jamie's School Dinners* and *Super Size Me* – there's been a sea change in public awareness about healthy eating, corporate power, workers' rights, industrial food production and all the other McLibel issues. These were niche subjects discussed on the fringes of society when we started in the mid-90s; now they are part of everyday mainstream discourse. Plus, McDonald's were forced to change many of their animal production methods for the better, their profits nosedived (although have since recovered) and, best of all, the UK government banned the advertising of junk food to children.

These real-world changes convinced me that I had stumbled into the perfect medium for maximizing the impact I could hope to make as a single individual. If you have a burning idea that you want to communicate – uncensored – with maximum possible emotional punch and a potential audience of tens of millions, you have to make a doc. Why? Because film is a mixed-media format – interviews, actuality, music, graphics – the size of the punch it packs can be so much bigger than single-media formats like books, songs, photographs or newspaper articles. Plus, the ninety-odd-minute length has been shown by the history of cinema to be the perfect time slot for people to follow, and become emotionally involved in, a story or an issue.

And, with the new distribution methods, you can now directly reach tens of millions of people without any editorial intervention from TV commissioners or from advertisers telling you to water down your message so they can sell more cat food in the ad breaks.

Despite finding the creative and collaborative process of filmmaking absolutely thrilling, I had no intention of jumping into bed with another story, as I found it impossible to imagine loving any other as I did *McLibel*. But then I read a newspaper headline over someone's shoulder on a train: 'Villagers in the Shadow of the Dam Prepare to Drown'. Without even knowing which country it was referring to (it turned out to be the Narmada dam in Jalsindhi, central India), I knew this would be my next film. Three years and many, many adventures – including a night in an Indian jail – later, my film *Drowned Out*, the true story of one family's inspired stand against the destruction of their land, homes and culture, sold to broadcasters round the world, was released in cinemas in the US and translated into many languages. *The Royal Gazette* in Bermuda said 'Documentaries rarely, if ever, come better than this'.

Throughout these years, I was becoming confident enough in my filmmaking skills to feel almost ready to tackle what was by then clearly the world's most important subject. But how? Eventually, with the help of a few beers, I cracked it: I would steal the structure of Steven Soderbergh's hit movie *Traffic*. His film followed the human stories of six people on all sides of the international drugs trade – corrupt cop, dodgy politician, middle-class drug user, etc. Nothing was straightforward, nobody was black and white, and everything was complicated by layers of conflicting desires and motivations. Mine would do the same for climate change, except documentary rather than fiction and real people instead of actors.

First problem: cash. I'd learned from *McLibel* that whoever owns the rights to the finished film controls the distribution. Say, for example, that the BBC had decided to commission and pay for my film back at the beginning, then it would have belonged to them. They'd have broadcast it on TV in the UK once or twice, possibly sold it to a couple of other countries, and that would have been it. No cinema release, no DVD, no festivals, no nothing. I sold *McLibel* to the WorldLink cable channel in the US for such a miniscule amount of money that the BBC would never have considered the deal, but – and this is the most enormous but – three million more people watched it.

By this stage, I had wildly ambitious plans for the new film, including animation sequences, orchestral scores and filming over three years in five countries. These things don't come cheap, even to practised DIY beg-borrow-and-steal filmmakers like ourselves. So the producer, Lizzie Gillett, and I sat around my flat with my dad one evening in 2004 and sketched out the details for what is now known as 'crowd-funding'. The resulting film, *The Age of Stupid*, ended up being funded by 350+ ordinary people and groups – including a hockey team and a women's health centre – who each invested between £500 and £35,000, making a total of £450,000 for the production and the same again for the distribution. Which may sound like a fair wodge of money to make a documentary, but is actually a fraction of what the costs should have been, as the 104-person crew all worked for about a third of their normal fees in exchange for a share of the profits. The crew and crowd-funders each own a slice of the *Stupid* pie and Lizzie and I have fun meeting up once a year to share out the cash to the funders, crew, production studios, animators and everyone else who helped along the way. Crowd-funding is now a hugely popular method of financing everything from books, bands and movies to community wind turbines and school solar panels. In 2004 the £900,000 we raised for *Stupid* was the largest amount raised for a film. There were all sorts of bonuses that came along with crowd-funding – 350 free PR agents, moral support in times of despair, country cottages in which to write the script – but by far the best thing about it is that it takes the decision-making about which films get made out of the hands of a few television executives. Now, if anyone has an idea for a film that they can convince a crowd is worth funding, they can go right ahead and make it. We are already beginning to see, in the UK at least, a marked improvement in the range of films that make it to the multiplexes. 'But didn't Al Gore already make the climate change documentary?' was a common question during the five years we spent making *The Age of Stupid*. It never failed to raise a weary smile. *Casablanca* had already done love, so why bother with *Brokeback Mountain*? *Apocalypse Now* did war, so what's the point of *Saving Private Ryan*? In my not-very-humble opinion, as the full horrors of climate change begin to unfold, love and war will become minor concerns, so we need more, not fewer, films about every aspect of the climate crisis and how we might yet solve it.

An Inconvenient Truth did the science. Fantastic. *The 11th Hour* analysed climate change alongside its non-identical twin, peak oil. *No Impact Man* got on to practical solutions from an individual's perspective and *The Power of Community* does the same at the community level. Our film focuses on the big moral human stuff.

In can-we-change-the-world terms, *The Age of Stupid* rocketed from the launch pad like nothing I'd ever known. At the 2009 premiere in a solar-powered cinema tent in London's Leicester Square (which was being broadcast live by satellite to 60+ cinemas round the country, making it the Guinness World Record-winning biggest ever premiere), the star of the film, actor Pete Postlethwaite, asked one of the assembled guests to join him on the stage. The lamb to the slaughter was Ed Miliband, who was then the government's Secretary of State for Climate Change. Ed thought he was in for a few gentle questions, but Pete launched his ambush: pulling out a giant pledge, Pete vowed never to vote for the Labour Party again – and to give his OBE back to the Queen – if Ed's department went ahead and commissioned a contentious new coal-fired power station. Ed was completely taken aback (see YouTube) and the newspapers lapped it up. Call it coincidence, but within a month the UK government made a major announcement: no new coal-fired power station will get consent unless it can capture and bury 25 per cent of the emissions it produces immediately – and 100 per cent of emissions by 2025 (when the technology will be better). A government source told the *Guardian* newspaper that this represented 'a complete rewrite of UK energy policy'.

The antics at the premiere catapulted the film into the public consciousness – which was handy, as we had no money to pay for advertising (the UK Film Council did some market research on our cinema goers and found that 98 per cent of them had come via word-of-mouth, an absolutely unheard of figure) – and the film even made it to number one at the UK box office, by screen average, in its week of release. Behind the scenes, we were frenziedly making as many international cinema and television deals as possible. The big audience numbers come from television broadcasts, when a couple of million people can watch at a single sitting, but in terms of inspiring action, it is often more powerful to have forty like-minded people sitting in a small room, following up the screening with a discussion. So we came up with a piece of software called 'Indie Screenings', to allow anyone anywhere to set up their own screening in their school, church or pub. Our natty algorithm works out an appropriate licence fee, after which, crucially, the organizers are free to charge for tickets and keep any profits for themselves. In the first six months, 1,100 local screenings were held and the system generated a very healthy £140,000 to go into the *Stupid* crowd-funding pot.

Throughout the film's production, I'd felt that the film itself was going to be my contribution to the battle and that, as soon as it was finished, I'd be free to rejoin my life. But as more and more people saw the film, the one question we were asked at ever-increasing volumes was: 'What can I do?' Walking through the park on the way to a public debate with Ed Miliband (strangely, he didn't speak to me again after the premiere ambush), I came up with the answer. 10:10.

10:10 (pronounced 'ten ten') asks individuals, businesses, schools, hospitals, the military, the monarchy – everybody – to commit to cutting their emissions by 10 per cent in a year.

Until this point, almost all emission reduction targets had been so far off as to be meaningless: 80 per cent by 2050? When are we going to start, 2049? And 10 per cent is in line both with what the science says is necessary and what is possible. I think that previous 'let's cut our carbon' campaigns have failed to capture the public's imagination because everyone makes the same calculation: what's the point of me changing my light-bulbs when Jimmy next door flies on holiday five times a year? 10:10 lets everyone who is concerned about climate change do something about it, in the context of many others across society doing the same.

10:10 launched at Tate Modern on 1 September 2009, and within forty-eight hours sign-ups included 10,000 individuals, 700 businesses, fifty schools, all the leaders of the main political parties, all the cabinet and the Prime Minister. As of 2012, the campaign has spread to forty-six countries and everyone from Colin Firth and Stella McCartney to London Zoo and the French Tennis Open – as well as the cities of Mexico, Paris and Marseille – are busy cutting their 10 per cent. One of the campaign's biggest successes to date is signing up 40 per cent of the UK's local councils – which means 25 million people in the UK will get their bins emptied and streets lit with 10 per cent less carbon than the year before.

So far, so thrilling. But the bigger aim of 10:10 is to prove to the world's politicians that society is ready for the new climate treaty our leaders so conspicuously failed to agree at the Copenhagen climate summit and Rio+20. A binding international treaty is now our only chance of doing what needs to be done: total global emissions must be stabilized by around 2015 and then must decrease very rapidly down to about 80 per cent of current levels by 2040 or so.

In 2021 we still do not have a treaty that will prevent our hitting the dreaded two degrees. Two degrees Celsius is thought to be the point at which we enter irreversible runaway climate change: two would lead to three, three to four, four to five, five to six... by the time we have heated our planet by an additional six degrees, it would be about over for most of life on Earth. In other words, our elected leaders have not addressed avoiding complete catastrophe. It is hard to think of a more complete failing of our political system.

As my biology teacher once said, if you put yeast into a jar with some sugar, they will gobble up the energy as quickly as possible, reproduce wildly out of control and then wipe themselves out in their own waste products. So far, our collective response to climate change has been of the yeast variety. 'We wouldn't be the first life form to wipe itself out,' says Pete Postlethwaite in *The Age of Stupid*, 'but what would be unique about us is that we did it knowingly.'

Further Reading & Useful Websites

McLibel by Franny Armstrong and Ken Loach
(1997, directors)

Drowned Out by Franny Armstrong (2002, director)

The Age of Stupid by Franny Armstrong
(2009, director)

www.spannerfilms.net
www.mcspotlight.org
www.1010global.org

BOB BROWN

Green Politician & Environmentalist, Australia

Born in Australia in 1944, Bob Brown has been a lifelong and courageous campaigner on a diverse range of issues, from gay rights to pacifism and environmentalism. In 1972, he moved to Tasmania, where he became a member of the United Tasmania Group, Australia's first-ever Green party, and director of the Tasmanian Wilderness Society. He was elected to the Senate in 1996 and in 2005 became parliamentary leader of the Australian Greens. He retired from the Senate in June 2012 to set up his foundation, www.bobbrown.org.au. The Bob Brown Foundation is a non-profit fund set up to support environmental campaigns in Australia and to assist activists who show 'real pluck and intelligence' in protecting ecosystems, species and wild and scenic heritage.

The Bob Brown Environmental Prize, launched in 2012, is an annual award to an individual or group who shows exemplary courage, wisdom, innovation or campaign skill and dedication to protect, enhance or highlight the environment of Australia and its region, including Antarctica.

I studied medicine at Sydney University and practised for twelve years after graduating in 1968, but I have always been an activist at heart. During my tenure at the Royal Canberra Hospital, I joined other doctors to certify young men who did not wish to fight in the Vietnam War as unfit to be conscripted. I moved to Tasmania in 1972 to take up a post as a GP and in the years after became active in the state's environmental movement, and joined the newly formed United Tasmania Group (the world's first-ever Green party). In 1976, I campaigned to overturn the law that made homosexuality a crime in Tasmania, and I spent a week fasting on top of Mount Wellington in protest against the arrival at Hobart of the nuclear-powered warship USS *Enterprise*.

Also in 1976, I helped establish the Wilderness Society, acting as director for five years from 1978, and so had to give up medical practice. In 1982–3 the society organized the blockade of the dam works on Tasmania's Franklin River. The blockade saw 1,500 people arrested for getting in the way of bulldozers building a huge rock-fill dam to block the Gordon and Franklin rivers and flood a vast area of rainforest for hydroelectricity. Six hundred people were jailed, including me. I spent nineteen days in Risdon Prison and on the day after my release in 1983, I was elected into Tasmania's Parliament.

In the streets of Hobart more than 20,000 people were protesting against the dam. In the parliament, all but two other members were voting for the dam and there was an air of great hostility. I soon found out that jail was a much friendlier place than parliament.

To cut a long story short, the popular unrest helped bring in a 'no-dams' government at the national election in

March 1983. The state and national governments resolved the issue in Australia's High Court, which ruled by four judges to three to uphold the World Heritage Convention and stop the dam.

Nowadays, this wild river wilderness attracts 200,000 visitors a year, creating thousands of jobs, and instead of receiving death threats from pro-dam locals, I am a welcomed visitor.

I often think back to the sunless days in my cell in Risdon Prison and say a quiet word of thanks to those feisty Tasmanian voters of 1983 who not only saved the Franklin River wilderness but, through me, gave an extra voice to the voteless myriad of our fellow creatures on this diversely magnificent planet of ours.

Since then, I have been assaulted and shot at during protests against logging at Tasmania's Farmhouse Creek in 1986 and jailed twice in 1995 for demonstrating to protect Tasmania's Tarkine Wilderness from logging. In 2006, I began a campaign of legal action to protect Tasmania's Wielangta forest. As a State MP, I introduced initiatives such as Freedom of Information, Death with Dignity, lower parliamentary salaries, gay law reform, banning the battery-hen industry, nuclear-free Tasmania and protection of native forests. Labor and Liberal voted against my 1987 bill to ban semi-automatic guns, seven years before the Port Arthur massacre.

In 1989, I led the five-member Greens parliamentary team, which held the balance of power with the Field Labor government. The Greens saved twenty-five schools from closure, created more than a thousand jobs through our local initiatives job scheme, doubled the size of Tasmania's Wilderness World Heritage Area to 1.4 million hectares, created the Douglas Apsley National Park, and supported tough fiscal measures to rid the state of the previous Liberal debt.

In 1996 I was elected to the Australian Senate, where I have since fostered national debates on climate change, Australia's involvement in war, the green economy, preventative healthcare, conservation, and human rights. I have introduced many private member's bills, including for electoral and parliamentary reform, rights of the territories, a ban on junk food advertising, ending mandatory sentencing of Aboriginal children, and to save Tasmania's magnificent wildlife-filled forests from logging. With Drew Hutton, I was instrumental in setting up the Australian Greens in 1992 and in 2005 was elected leader. The federal Greens parliamentary team expanded to five in 2007 and ten in 2010. In 2011, this Greens team led the passage of world-leading legislation to cut greenhouse gas pollution and foster renewable energy in this lucky, sunny country of ours.

I have an holistic view of life on planet Earth – we are all in this together. At the fortieth anniversary celebration of the world's first Green party meeting, which had taken place in Hobart Town Hall on 23 March 1972, I outlined a way forward. It drew howls of rage from Australian media. Here it is.

Never before has the Universe unfolded such a flower as our collective human intelligence – so far as we know. Nor has such a one-and-only brilliance in the Universe stood at the brink of extinction, so far as we know. We people of the Earth exist because our potential was there in the Big Bang, 13.7 billion years ago, as the Universe exploded into being. So far, it seems like we are the lone thinkers in this vast, expanding Universe. However, recent astronomy tells us that there are trillions of other planets circling sun-like stars in the immensity of the Universe, millions of them friendly to life. So why has no one from elsewhere in the cosmos contacted us?

Surely some people-like animals have evolved elsewhere. Surely we are not, in this crowded reality of countless other similar planets, the only thinking beings to have turned up. Most unlikely! So why isn't life out there contacting us? Why aren't the intergalactic phones ringing?

Here is one sobering possibility for our isolation: maybe life has often evolved to intelligence on other planets with biospheres and every time that intelligence, when it became able to alter its environment, did so with catastrophic consequences. Maybe we have had many predecessors in the cosmos but all have brought about their own downfall.

That's why they are not communicating with Earth. They have 'extincted' themselves. They have come and gone. And now it's our turn.

Whatever has happened in other worlds, here we are on Earth altering this bountiful biosphere, which has nurtured us from newt to Newton.

Unlike the hapless dinosaurs, which went to utter destruction when a rocky asteroid plunged into Earth sixty-five million years ago, this accelerating catastrophe is of our own making.

So, just as we are causing that destruction, we could be fostering its reversal. Indeed, nothing will save us from ourselves but ourselves.

We need a strategy. We need action based on the reality that this is our own responsibility – everyone's responsibility.

So democracy – ensuring that everyone is involved in deciding Earth's future – is the key to success. For comprehensive Earth action, an all-of-the-Earth representative democracy is required. That is, a global parliament.

In the Gettysburg Address of 1863, Abraham Lincoln proclaimed: 'We here highly resolve … that government of the people, by the people, and for the people shall not perish from the earth.'

For those who oppose global democracy, the challenge is clear: how else would you manage human affairs in this new century of global community, global communications and shared global destiny?

One evening, when I got back to bed at Liffey after ruminating under the stars for hours on this question, [my partner] Paul enquired, 'Did you see a comet?' 'Yes,' I replied, 'and it is called "Global Democracy".'

A molten rock from space destroyed most life on the planet those sixty-five million years ago. Let us have the comet of global democracy save life on Earth this time. Nineteen years ago, after the invasion of Iraq, which US President George W. Bush ordered to promote democracy over tyranny, I proposed to the Australian Senate a means of expanding democracy without invasion. Let Australia take the lead in peacefully establishing a global parliament. I explained that this ultimate democracy would decide international issues. I had in mind nuclear proliferation, international financial transactions and the plight of our one billion fellow human beings living in abject poverty.

In 2003, our other Greens Senator, Kerry Nettle, seconded the motion, but we failed to attract a single other vote in the seventy-six-seat chamber. The four other parties – the Liberals, the Nationals, Labor and the Democrats – voted 'No!'. As he crossed the floor to join the 'no's, another senator called to me, 'Bob, don't you know how many Chinese there are?'

Well, yes, I did. Surely that is the point. There are just 23 million Australians among seven billion equal Earthians. Unless and until we each accord equal regard to every other citizen of the planet – friend or foe, regardless of race, gender, ideology or other characteristic – we, like them, can have no assured future.

The Athenians 2,500 years ago and the British 180 years ago gave the vote to all men of means. After Gettysburg, the US made the vote available to all men, regardless of means. One man, one vote.

But what about women, Louisa Lawson asked in 1889: 'Pray, why should one half of the world govern the other half?'

So in New Zealand in 1893, followed by South Australia in 1895 and the new Commonwealth of Australia in 1901, universal suffrage – the equal vote for women a well as men – was achieved.

In this second decade of the twenty-first century, most people on Earth get to vote in their own countries. Corruption and rigging remain commonplace, but the world believes in democracy. As Winston Churchill observed in 1947: 'Many forms of government have been tried in this world of sin and woe. No one pretends that democracy is perfect or all-wise. Indeed, it has been said that democracy is the worst form of government except all those other forms that have been tried from time to time.'

Yet in Australia and other peaceful places that have long enjoyed domestic democracy, establishing a global democracy – the ultimate goal of any real democrat – is not on the public agenda.

Exxon, Coca-Cola, BHP Billiton and News Corporation have much more say in organizing the global agenda than the five billion mature-age voters without a ballot box.

Plutocracy, rule by the wealthy, is democracy's most insidious rival. It is served by plutolatry, the worship of wealth, which has become the world's prevailing religion. But on a finite planet, the rule of the rich must inevitably rely on guns rather than the ballot box – though, I hasten to add, wealth does not deny a good heart.

We instinctively know that democracy is the only vehicle for creating a fair, global society in which freedom will abound, but the extremes of gluttony and poverty will not. Mahatma Ghandhi commented that the world has enough for everyone's need but not for everyone's greed.

So what's it to be: democracy or guns? I plunk for democracy.

The concept of world democracy goes back centuries but, since 2007, there has been a new movement towards an elected, representative assembly at the United Nations, in parallel with the unelected, appointed General Assembly. This elected assembly would have none of the General Assembly's powers but would be an important step along the way to a future, popularly elected and agreeably empowered global assembly.

We Earthians can develop rosier prospects. We have been to the moon. We have landed eyes and ears on Mars. We are discovering planets only hundreds of light years away, which are ripe for life. We are on a journey to endless wonder in the cosmos and to realizing our own remarkable potential. To give this vision security, we must get our own planet in order.

The political debate of the twentieth century was polarized between capitalism and communism. It was about control of the economy in the narrow sense of material goods and money. A free market versus state control. Bitter experience tells us that the best outcome is neither, but some of both. The role of democracy in the nation state has been to calibrate that balance.

In this twenty-first century, the political debate is moving to a new arena. It is about whether we expend Earth's natural capital as our population grows to ten billion people in the decades ahead with average consumption also growing.

We have to manage the terrifying facts that Earth's citizenry is already using 120 per cent of the planet's productivity capacity – its renewable living resources; that the last decade was the hottest in the last 1,300 years, if not the last 9,000 years; that we are extincting our fellow species faster than ever before in human history; and that to accommodate ten billion people at

American, European or Australasian rates of consumption, we will need two more planets to exploit within a few decades.

It may be that the Earth's biosphere cannot tolerate ten billion of us big consuming mammals later this century. Or it may be that, given adroit and agreeable global management, it can. It's up to us.

Once more, the answer lies between the poles: between the narrow interests of the mega-rich and a surrender to the nihilist idea that the planet would be better off without us.

It will be global democracy's challenge to find the equator between those poles, and it is that equator that the Greens are best placed to reach.

One great difference between the old politics and Green politics is the overarching question that predicates all Green political decisions: 'Will people one hundred years from now thank us?'

In thinking one hundred years ahead, we set our community's course for one hundred thousand years: that humanity will not perish at its own hand but will look back upon its twenty-first century ancestry with gratitude.

When the future smiles, we can smile too.

That query, 'Will people a hundred years from now thank us?', should be inscribed across the door of Earth's parliament.

So let us resolve that there should be established for the prevalence and happiness of humankind a representative assembly a global parliament for the people of the Earth based on the principle of one person one vote one value; and to enable this outcome that it should be a bicameral parliament with its house of review having equal representation elected from every nation.

An Earth parliament for all. But what would be its commission? Here are four goals: Economy, Equality, Ecology and Eternity.

To begin with, economy – because that word means managing our household. The parliament would employ prudent resource management to put an end to waste and to better share Earth's plenitude. For example, it might cut the trillions of dollars of annual spending on armaments. A cut of just 10 per cent would free up the money to guarantee that every child on the planet has access to clean water and enough food, as well as a school to attend to develop his or her best potential.

The second goal is equality. This begins with equality of opportunity – as in every child being assured of that school, where lessons are in his or her own first language, and a health clinic to attend. Equality would ensure, through the fair regulation of free enterprise, each citizen's well-being, including the right to work, to innovate, to enjoy creativity and to understand and experience and contribute to defending the beauty of Earth's biosphere.

Which brings me to the third goal: ecology. Ecological well-being must underpin all outcomes, so as to actively protect the planet's biodiversity and living ecosystems. 'In wildness,' wrote Thoreau, 'is the preservation of the world.' Wild nature is our cradle and the most vital source for our spiritual and physical well-being. Yet it is the world's most rapidly disappearing resource.

So I pay tribute to Miranda Gibson, who sat for months 60 metres high on her tall tree platform, even as the rain and snow fell across central Tasmania. She defied the loggers coming to destroy her ancient tree. In Miranda's spirit is the saving of the world.

And lastly, eternity. Eternity is for as long as we could be. It means beyond our own experience. It also means 'for ever', if there is no inevitable end to life. Let's take the idea of eternity and make it our own business.

I have never met a person in whom I did not see myself reflected. Some grew old and died, and I am now part of their ongoing presence on Earth. Others have a youthful vitality, which I have lost and will soon give up altogether.

These youngsters will in turn keep my candle – and yours, if you are aged like me – alight in the cosmos. In this stream of life, where birth and death are our common lot, the replenishment of mankind lights up our own existences. May it go on and on and on...

The pursuit of eternity is no longer the prerogative of the gods: it is the business of us all, here and now.

Drawing on the best of our character, Earth's community of people is on the threshold of a brilliant new career in togetherness. But we, all together, have to open the door to that future using the powerful key of global democracy.

I think we are intelligent enough to get there. My faith is in the collective nous and caring of humanity, and in our innate optimism. Even in its grimmest history, the optimism of humanity has been its greatest power. We must defy pessimism, as well as the idea that there is any one of us who cannot turn a successful hand to improving Earth's future prospects.

I am an optimist. I'm also an opsimath. I learn as I get older. And I have never been happier in my life. Hurtling to death, I am alive and loving being Green.

I look forward in my remaining years to helping spread a contagion of confidence that, together, we people of Earth will secure a great future. We can and will retrieve Earth's biosphere. We will steady ourselves – this unfolding flower of intelligence in the Universe – for the long, shared and wondrous journey into the enticing centuries ahead.

Let us determine to bring ourselves together, settle our differences, and shape and realize our common dream for this joyride into the future. In that pursuit, let us create a global democracy and parliament under the grand idea of one planet, one person, one vote, one value, one planet.

We must, we can, we will.

Further Reading & Useful Websites

Memo for a Saner World by Bob Brown (2004)

Planet Earth: Inspirations and thoughts from a planet warrior by Bob Brown (2019)

www.greens.org.au
www.bobbrown.org.au

HELEN BEYNON

Environmental Activist, UK

Helen Beynon is the pen name of Helen Baczkowska, a writer, ecologist and environmental activist who lives and works in rural Norfolk. Helen's book, *Twyford Rising – Land and Resistance*, is an oral history of Britain's first direct-action road protest at Twyford Down in Hampshire. Those protests are widely seen as part of the foundations of the current environmental movement. Helen has also collaborated on an Arts Council funded project recording places and communities impacted by fracking in the UK and is currently working on a book about common land in Britain today.

On a May afternoon in 1992, I walked on Twyford Down for the first time. Two friends and I followed a footpath from the centre of Winchester, the county town of Hampshire and once, long ago, the capital of England. We walked beside small streams that ran swiftly over pale stones, crossed the damp, dark soil of meadows beside the river Itchen and then entered the broad sweep of Plague Pits Valley. This dry curving valley of grassland and scattered scrub was, we were later told, where the plague dead of Winchester had been buried; now it was nature reserve, grazed by a few black and white Gypsy ponies. To the north was St. Catherine's Hill, topped by a clump of beech trees and an Iron Age hillfort. To the south and east were steep slopes up to the chalk hill of Twyford Down. At the top of the down, the landscape was rutted with ancient trackways, some up to thirty feet deep and interlinked, like a secret network of small valleys. These 'holloways' had been made by the footsteps of medieval drovers, travellers, farmers, cattle and sheep; over time the paths, known as the 'Dongas', were washed deeper by rainfall. The name, it was said, had been bestowed by a well-travelled Winchester school master, who had seen similar sunken roadways called 'dongas' in Africa.

On the day that my friends and I first walked onto the down, a small stage with speakers was laid out on the short, rabbit-grazed turf. There was music and drumming, but I wasn't paying much attention. I was too busy falling in love with the land. The ruts of the Dongas and the steep slopes were too difficult to plough and here, among the hawthorn bushes, chalk grassland thrived. The sward was so full of life that I slipped off my boots and felt the warm grass beneath my bare feet. To walk on this land in heavy footwear seemed wrong, sacrilegious perhaps. I lay down, entranced, for among the grasses were orchids almost ready to bloom, flowers of wild thyme, cowslip and pale-yellow rockrose. Everywhere there were tiny black spiders, ants and bumblebees. Eventually, the speakers stepped up the mike, then a crackling satellite 'phone connected us with delegates at the United Nations Earth Summit in Rio do Janiero. Together we shouted a message, all the way from the sunshine of an English summer day to Brazil: 'Stop the destruction of Twyford Down'.

The hill, the curious, rippling landforms of the Dongas gullies, the grassland teeming with insects and part of the water meadows beside the river were all poised to be destroyed for a motorway. Tonne after tonne of chalk, we were told by the speakers, would be removed, until the high down itself was almost level with the river. The resulting cutting through the hill would be half a mile wide and nearly as deep. The excavated chalk would be spread on the meadows below, creating a causeway for the road to pass over the river and across a seventeenth-century canal that linked Winchester with the seaport of Southampton. This was one of the most protected landscapes in southern England and part of the South Downs Area of Outstanding Natural Beauty. The downland and the riverside meadows were designated as Sites of Special Scientific Interest, in recognition of their value for wildlife, while the Dongas trackways and a cluster of Bronze Age burial mounds were Scheduled Ancient Monuments. These designations, and the opinions of the Winchester citizens who walked on the down and loved its peaceful beauty, meant nothing once central government decided to extend the M3 motorway and shave three minutes off the journey from London to Southampton.

As the speeches finished, someone grabbed the microphone:

'Down there', he shouted and pointed. 'Down there, they are right now filling in the canal. Let's go an' stop them.'

A hundred or so people skittered down the hillside and waded into water that swirled white with the chalk that was being bulldozed into a dam. We pulled the bank apart with our bare hands and created waves to wash some of it away. We stopped them for that day at least. I was twenty-three years old and my life had just changed forever. I had grown up watching news clips of Greenpeace stopping whaling boats, acutely aware that one day I wanted to intervene in the destruction of wild creatures and places. At the same time, I longed to know the names of every plant in the hedgerow or road verge; my parents, unable to answer my questions, did the best thing they could have done and bought me a wildflower book. I taught myself to identify cowslips and cow parsley, to tell the difference between the tall, fine-leaved meadow buttercup, which often grows on old pastures, and the creeping buttercup common in gardens and on footpaths. Eventually I trained as an ecologist, but the ways into environmental campaigning seemed harder – most organizations used only a few, skilled activists and many of their actions were, I discovered, carefully orchestrated press stunts. I was happy to contribute to their efforts, but something more radical, more direct, felt essential. For a while, I was a hunt saboteur – an experience that made me a good map reader, skilled in using shortwave radios and thinking quickly. It was worthwhile, but the continued extinction of species and habitats gnawed always at my mind. Then, in the late 1980s, I read about a new group called Earth First! sitting in trees and blockading lorries to stop the logging of old-growth forests in the United States. I felt that change was on its way and that the environmental movement would, before long, become bolder and more visible, responding to the urgency being felt by increasing numbers of people across the globe.

Falling in love with the chalk downland and the dam-busting that followed was the beginning of five years of my life spent on road protests and a lifetime of campaigning that continues today. Twenty years after that sunny afternoon, fellow campaigner Chris Gillham and I began collecting the stories of some of those who had taken part in the Twyford protests into a book, *Twyford Rising – Land and Resistance*. Capturing the story seemed essential: this was the first place in Britain where people came together to take environmental direct action on this scale, standing in the way of construction work, sitting on vast, earth-moving machinery and squatting land on the route of the road. Twyford Down is a part of the foundations of the modern environmental movement, where people are prepared to state that some issues are so important we will act even if it risks arrest or imprisonment. If, we reasoned, the law cannot protect places like Twyford Down, and if governments can act beyond the law by destroying places that are supposed to be protected, then why should we respect the rules?

Now, when I give readings from the book, certain questions come up again and again: how do you as an individual find hope, or keep going as an activist? What part did the deliberate damage of construction machinery play and was it a good approach? What can the past tell us about what tactics work, how people become motivated, what makes politicians or businesses listen? The readings always finish with a wide-ranging discussion, a reminder that no one person ever has all the answers to these questions, but by working together we might just find them. I am, of course, also a writer and we are often accused of being nosy. I prefer to say that I am fascinated by the stories of others, especially unheard voices, including those of the land and the non-human lives that inhabit it. From the very start, it was important to me to capture the words and memories of as many of those who were part of the Twyford Down protests as I could, emphasizing the diversity of experience, people and ideas.

In the summer of 1992, I was living near Twyford Down and started to join the protests when I could. I quickly learned that there were three main groups driving the campaign – the local residents, who had opposed the road over decades and who were still attempting to stop it through the European courts; the fledgling UK Earth First! movement, dedicated to direct action under the slogan 'No Compromise in Defence of the Earth' and a small group living on the down, who had adopted the name of the land and were known as the 'Dongas Tribe'. The connection between local opponents of the cutting and Earth First! had begun in January that year, just as work was scheduled to start. The first stage of construction was the demolition of two disused railway bridges on the route of the road and, realizing the urgency of the hour, one of the local residents, a former Conservation Party councillor, decided to act. David Croker had, like others, opposed the road through years of planning applications, public inquiries and judicial review. With work about to commence, he sought out a meeting of an Earth First! group at a squat in Brixton, laid the facts before them and asked for their help.

Up to that point, Earth First! was very much in its infancy in the UK, with a handful of activists mostly organizing blockades of timber being imported from North America or the South Pacific. The group answered the call to Twyford, occupied one of the bridges overnight and stopped a bulldozer by locking someone to it by the neck with a bicycle 'D-lock' – the first time such tactics had been used in Britain.

In subsequent weeks, Earth First! decided to devote as much effort as possible to getting people to the construction site – in those days before social media, mobile phones or even email, this meant handing out leaflets at gigs and festivals and giving talks on university campuses. Despite the busy organizing activities, Earth First! existed as little more than a banner to gather beneath. In the UK, a network of small, local groups evolved and operated without committees or leaders. This was a deliberate strategy to avoid the collapse that can happen in organizations if leaders leave or are imprisoned, but it also ensures consensus and inclusivity. Alex was one of the women who lived on the Donga's camp and was often seen playing a tin whistle as we occupied worksites, her face painted with colourful spirals. In my book, she talks of how Earth First! 'has no law, no hard and fast policies', but 'exists through the people who actively make it up at any given time.'

The third strand of the protest, the Dongas Tribe, began in the spring of 1992 as a camp that shifted a few times and eventually nestled among the thorn bushes and old trackways on the down. The camp was much more than a place for visiting activists to stay and the mostly young people who lived there full time evolved a close connection to the land that was infused with myth, music, art and sometimes with spirituality. One protestor, Larf, went to live on the down after seeing an Earth First! newsletter and explained: 'My understanding is that for many of us the initial protests were about preserving the land and protecting the Earth, the road was incidental. The protest would still have happened if the plan was to build a housing estate on Twyford Down.'

Colin was a musician who joined the camp and he too spoke of being motivated by much more than just the destruction of the down:

'It's to do with lifestyle, one that works. This one's going to strangle itself slowly unless you do everything you do with respect for the land that's supporting you. '

Opposing motorways was also a challenge to the freedom most people have grown up with – to travel, to own cars, to see them as a necessity, not a luxury. It was a battle not just against an individual place being trashed, but about a whole transport system that was slipping out of control.

The people who lived on the land, in those days before mobile phones, livestreams and instant messaging, developed a close-knit community. They sat around the fire in the chill of an evening, playing music, telling stories and exploring ideas. Many of us, campers or visitors, had grown up hearing about acid rain, the hole in the ozone layer and the ever-present shadow of nuclear war. Now we talked about the links between unbridled capitalism and climate change, fossil fuels and extinction and how we might find ways to reverse the most destructive tendencies of our species.

At night, on the quiet land above the city, the Tribe slept in 'benders' – sturdy shelters of tarpaulin over bowed and freshly cut hazel rods. Benders were once used by Gypsies, but were also a feature of the women's peace camps established outside Greenham Common military base in the early 1980s. The Greenham camps were born of a very real concern that the nuclear weapons of mass destruction stockpiled by the Soviet Union, the United States and their allies could easily obliterate most life on Earth several times over. By 1992, the camps had been present for over ten years. They bore witness to the ground-launched missiles at Greenham Common and the women frequently interrupted military 'exercises', or cut the airbase fence and ventured inside the compound. In the early days of the Twyford protests, some of the women from the camp joined us and passed on their knowledge of direct action, court and arrest. Their wisdom, courage and even their songs were quickly woven into our own dissent. Sitting on vast yellow diggers, we would sing, as they did, 'you can't kill the spirit; she is like a mountain, old and strong. She goes on and on'.

Ultimately, and in spite of the protests, the cutting through Twyford Down was built, the canal was dammed and part of the Itchen river put into a pipe. Now there are noise and fumes where once there was sunlight and wildlife, a land where the people of Winchester had walked for centuries and where ancient people had lived or buried their dead. In the months before it opened, there were frequent attempts to stop work and the brutal break-up of the camp on the Dongas by security guards. This was the first time in Britain that such a force had been used against environmental protestors. The local constabulary stood by, watching as people were physically and sexually assaulted by the Department of Transport's hired men and, apart from a solitary *Guardian* journalist, the national media largely ignored us for over a year. In those months of almost daily direct action, arrests were frequent and a High Court injunction resulted in people being sent to prison, or living for years with the threat of losing their homes as the government tried to sue for millions of pounds in lost time, extra security and damage to machinery.

Emma Must, one of those imprisoned in the summer of 1993, was in her twenties when she addressed the judge from the dock in the High Court, knowing she would soon be going to a cell in Holloway prison: 'I grew up near Twyford Down and walked over its beautiful rolling slopes on many occasions. Now Twyford Down is a gaping chasm and the Itchen valley, at its foot, has an enormous embankment stretching across it.'

Emma, and the others sent to prison for defying the injunction, also used this opportunity to draw attention to the government's then road-building programme, with its threats to hundreds of homes, other protected historic sites and greenspaces. Her court statement continued: 'I cannot consent to such wilful annihilation of such an immense swathe of our heritage by an ephemeral government. To have ceased protesting in the face of this desecration would have meant giving that consent.'

As I gathered these statements and memories into *Twyford Rising*, I felt I was revisiting some of the fundamental questions of environmental activism and reminding myself how important these are. How do we live with minimal harm to nature? Is honouring the Earth as a living being, or as a wonderful, fragile creation, one way of stopping the damage that our species carries out daily? How does it feel to be part of a community founded on a common cause? And as, of course, not everyone agreed all the time, how does that community resolve its differences and stay strong? The people I interviewed also spoke about why direct action was needed and how they dealt with their first arrest. Debs, another resident of the Dongas camp, speaks of direct action as 'looking at authority and saying…we are going to play it our way, 'cos your way is crap and it hasn't done any good'. Colin explained how people realized that getting arrested 'wasn't enough to change your feeling of doing what needed doing'.

Beyond actions and arrest, days on the Dongas were taken up with making clothes, learning instruments or tunes and telling stories. We saw ourselves as heirs to a long tradition of rebellion and taught ourselves an alternative British history to the one that focuses on empire and monarchy. I had been lucky enough to grow up hearing some of these tales as told by my family and realize that my activism is motivated by social justice as much as it is by my love for the natural world. My great-grandfather, Dan Beynon, was sent to work in a coalmine in South Wales when he was eight years old. With only a flickering candle against the dark, he opened and closed wooden doors for the pit ponies and their rattling trucks to pass through. Somehow, he taught himself to read and write, first in Welsh and then in English. When he grew up he helped to found a mine workers' union and spent a lifetime organizing strikes and supporting calls for paid weekends, paid holidays and sick leave – working conditions we now see as basic rights. Although I never met Dan, his children and grandchildren inherited his passion, so that discussions of politics and injustice dominated family gatherings. Perhaps that is one reason why I see activism not as a negative act, or as being 'against something', but as faith that collective action can improve the ways we live with each other and on the Earth.

Dan's struggles seemed very close to home when I was a teenager, growing up in the English Midlands, with the year-long Miner's Strike on our doorstep. Fifty or more police riot vans frequently parked in our village on the way to the picket lines and there were road blocks at nights on country lanes. Like many of our neighbours, we donated food and hand-me-down clothes to the striking families when we could. It is easy to forget that in Britain, the 1980s were a bleak decade to grow up in – economic recession and unemployment, communities pushed to the brink by the closures of mines, steelworks, factories. The feeling that society, Nature, anything not centred on money or profit, was without value. And always, the threat of nuclear war hung over us, so much so that my mum filled a cupboard with tins of food just in case. One day, like so many others, she realized she could no longer live with the fear, joined a local women's peace group and travelled to Greenham Common for demonstrations.

By the time the Twyford Down cutting was opened in July 1994, a network of protest camps had been established across Britain. These were not just where new roads were planned, but at quarries, opencast coalmines and later at airports; the places under threat were not only wild land, but streets and houses and city parks. At the same time, some of those who had lived on the Dongas were trying their hand at a car-free travelling lifestyle, with horse-drawn wagons, hand carts and bicycle trailers. A shifting, drifting community grew up between all of the spaces we occupied and the scope of the protests widened too, with street parties shutting down roads and showing what could happen if towns were freed from the tyranny of traffic. Days of action were carried out against genetically modified crops, on peat moors being dug up for garden compost and in support of people around the world fighting for their environment. I remember shutting down petrol stations and fuel depots in solidarity with the Ogonis' struggle to get Shell to clean up the damage caused by oil spills in Nigeria and occupying the offices of companies profiting from the destruction of forests for timber or mining. I travelled to Canada to join First Nation activists opposing the industrial logging of their homeland. In the mountains where we stayed, I watched bear cubs playing by a river and heard wolves howling in the forest in the evening. In Iceland, I was one of many campaigners from across the world to stand beside farmers opposing huge hydro-electric power stations, which were being constructed for aluminium smelting as part of the USA's carbon trading policies. The dams, some of which have since been scrapped, would flood high, rocky valleys, where flocks of pink-footed geese came every summer to rear their young. While in Iceland, I met people from South America and India, where similar schemes were internationally funded; knowledge of this and the punishing repayment of national debts, led to the organization of mass demonstrations at meetings of the G8 groups of countries, the World Bank and the International Monetary Fund.

In *Twyford Rising*, Phoenix talks of a movement that 'spiralled out from the ancient capital of Winchester'. He says that Twyford inspired a generation to 'rise up and resist', seeding the Climate Justice movement, the global Occupy camps and anti-fracking movement in the UK. It is a bold claim and more likely, the Twyford protests were part of a change in consciousness occurring around the world. Yet, for those of us involved at time, Twyford Down gave us the chance to fight back, to make a stand and help us reclaim our space in the world. It gave us a place to rally to, to defend and to be the centre of our struggle; somewhere for those of who us who felt so dispossessed by the world we had grown up in could be together around the fire, beneath the stars, sowing the seeds of a better way of living.

When people ask me, as they often do, if it feels like a failure that the cutting was finally built, I always pause. It takes me a moment to find the beginning of a long answer. After nearly thirty years, I still feel grief for the loss of the down and the tranquil water meadows below. Yet, other roads were stopped because of what was started at Twyford and direct action is now, across the world, part of how we speak out for all of the lives that live on this planet beside us. That feels something like success, even if bittersweet. For me, some of the most remarkable stories in *Twyford Rising* are those of people who overcame their fear of arrest and authority. Environmental direct action has been my life for thirty years, but it is a big step to take and, for many different reasons, it is not for everyone. Any movement for change needs a broad spectrum of people working at different levels, from researchers, legal experts and engineers to lobbyists and policy makers. The trick, in my view, is to find a way to recognize that continuum and the variety of roles within it.

As I came to the end of writing *Twyford Rising*, I realized I was facing a dilemma common to many writers: how to finish the book? What was apparent to me, as I interviewed people, was that the story had not ended and that most of those involved had carried those years with them all their lives. Consequently, I closed the book with reflections from people who have set up housing or workers co-operatives, continued campaigning and tried their hands at low-impact lifestyles. Activism remains part of my own life and is still mostly motivated by my love of the natural world. Not too far from where I live, a new road is being planned, crossing a chalk river that reminds me of the Itchen near Twyford, threatening to obliterate woodlands and a colony of rare bats.

As I write these words, summer is turning into autumn and bright orange pumpkins are ripening in my vegetable garden; the window by my desk is open and two young buzzards, not long fledged from their nest, are wheeling and calling, high on a thermal above. The older I get, the more I feel grateful that I have my garden and encounters with Nature to sustain me in the ongoing fight for the natural world and our planet.

Further Reading & Useful Websites

Twyford Rising: Land and Resistance by Helen Beynon with Chris Gillham (2020)

DEEPAK CHOPRA

Mind–body Healing Pioneer, USA

Born in India in 1946, Deepak Chopra is a doctor, pioneer in integrative medicine and a globally acclaimed writer of over ninety spiritual and self-help books. His book *Ageless Body, Timeless Mind*, published in 1993, sparked a revolution in Western attitudes to complementary medicine, which recognizes the importance of the mind/body/spirit connection in healing illness and promoting good health.

In 2002, Deepak founded The Chopra Foundation, a non-profit organization dedicated to improving health and well-being. He also founded Chopra Global, a modern-day health company at the intersection of science and spirituality.

I was raised in post-war India, a newly independent nation whose social and cultural values were slowly being transformed. My father, Krishan Chopra, was a famous physician, a cardiologist who gave me his passion for knowledge and service; my mother, Pushpa, was a humanitarian who taught me to respect life. Living in India, my family embraced both Eastern and Western influences. I grew up in the India of Gandhi's legacy. A great modern nation was being born, but it was integrating into its culture all the traditional beliefs. It was a country in search of an identity. I witnessed extreme poverty and despair, while the foundation of a technological revolution was being built. The once-rigid caste system was breaking down among educated people, but it still existed, and I heard stories about members of the Brahmin class returning to their home to take a bath because they had been touched by the shadow of someone from the lowest caste. While arranged marriages were still common, more young Indians were making their own choice of a partner; yet the servant who attended us throughout our entire childhood had been given to our father as part of our mother's dowry. And while my brother and I attended mainly Irish-Christian schools, we were raised Hindu. I studied science in school, and at home learned about our mother's deceased brother, who the family believed had been reincarnated. When he was only four years old, he could recite long passages from the sacred text, the Vedas; he gave up his pleasures to the poor and needy, tried to borrow rupees to pay back a debt that he explained was owed to a servant in a previous life, and finally, predicted his own death.

I remember in Poona, India, 1954, I was seven years old. My mother was away, visiting my grandmother's house, and had taken my four-year-old brother, Sanjiv, with her. I was home alone after school until my father came in around six o'clock. He looked excited and was beaming. He said, 'Hurry up! Wash your face, change your clothes, put on a tie, we're going out for some fun.' As an inquisitive child, I asked, 'Where are we going?' And he responded, 'No questions, wait and see. We're going to have the time of our lives!' Immediately, I changed my clothes, put on my school tie, and hurried out onto the veranda, where my father waited impatiently with his bicycle. I jumped onto the handlebars and off he pedalled through the warm summer night. I had no idea where we were going, but I was thrilled all the same.

As we got nearer to our destination, it became clear that we were headed for the army barracks, then for a small brick building known as the MI room, which was a combined clinic and dispensary for the British Army. Officers, common soldiers and their families all came there with their sore throats, aches and pains and broken bones. My father worked there, but I had no interest in it at all. The MI room also held a makeshift lecture hall that could seat about seventy-five people.

When we arrived, there were scores of bicycles parked in a jumble outside the facility. The hall was packed and bursting with military doctors and nurses and smelled of starched khaki and crisp white coats, which were everywhere. The air buzzed with excitement, and naturally there was not a seat to be had. What was this all about? In that charged atmosphere, it was only natural that a little boy should feel especially excited. We all waited.

Then the applause broke out. People kept clapping harder and jumping in front of me. The object of this clamour turned out to be an old, bespectacled, European-looking gentleman, who slowly approached and mounted the stage. My memory tells me that he had a white beard, but it may be playing tricks, since I outfitted all old, European-looking gentlemen with one. He took the slightest bow of acknowledgement to his ovation, which in time died away. The lights went out, and the hall became very quiet. The European gentleman began to talk and at the same time started projecting pictures on a screen by his side.

Of course, I couldn't understand a word he was saying, and all his pictures showed the same thing: lots of white dots with haloes around them.

I asked my father, 'What's he doing, Daddy?' 'Hush, son,' was all he could say. 'I will explain it to you later.'

We got home very late and my father tucked me into bed with a gleam in his eyes. He kept me awake one last minute and told me that the European gentleman had been describing the discovery of penicillin. That man that night was Sir Alexander Fleming.

The night my father took me to see Fleming, I had such a vivid dream that I can still remember it. I saw thousands of soldiers fallen on a battlefield, all dead. A kindly European man wearing spectacles and a beard was walking among the bodies, spraying them from a nozzle with little white dots. And as he sprayed the men, each one got up and walked away. Gradually everyone was sprayed, and not a single body remained. But right in the middle of the field lay a black cobra in a pool of its own blood. Its head was smashed and its neck broken. A small mongoose leaped around it in fits and jumps. The European approached and sprayed the snake, but nothing happened. He kept spraying, more and more, until he became frantic, but the cobra still lay in its pool of blood.

A very obscure dream for a small boy – or was it? I have decided to remember it, across the space of fifty-five years, as a story about medicine and me.

While the nation was testing its independence, the British colonial influence remained extremely strong. Our father, a famed cardiologist, who was trained in the British medical system, eventually became a member of the Royal College of Physicians, and was even introduced to Sir Alexander Fleming. But like most educated Indians, Krishan Chopra remained ambivalent about the British. When he arrived in London, for example, the very first thing he did was find a shoeshine boy, so he could sit upon a high stool and look down upon an Englishman shining his shoes.

Krishan and Pushpa Chopra sculpted my life. They were extraordinary people. In the Second World War, Krishan served in the Indian Army and was captured by the Japanese. He watched in horror as the Japanese darkened the faces of his British comrades with shoe-polish then shot them dead, sparing him because he wasn't a white man. His nightmares would haunt him for the rest of his life, and that experience would reinforce his determination to prolong life. After the war, his military assignments each lasted three years, after which he was transferred. He became so beloved that I vividly remember standing on a train platform, waiting for the train that would take him to his next assignment, while thousands of people turned out to honour him.

His military assignments caused our family to move throughout India, allowing us to experience the expanse of our country. And when father started his private practice, he treated anyone who came to him, accepting whatever they could afford as payment.

Wherever we lived, our mother, Pushpa, was our immovable anchor. When thieves broke into the house and beat the servants, Pushpa protected them, offering herself as a replacement hostage. After she handed the thieves her jewellery, one of the thieves handed back her earrings, telling her they belonged on her, then kissed her feet, apologized and asked for her blessing. Her aura seemed magnetic; and we were told about the day India's charismatic first Prime Minister, Jawaharlal Nehru, visited the city where my parents were stationed. Pushpa had told her friends that the Prime Minister would stop to greet her, a seemingly impossible boast. Vast crowds lined the streets to welcome him, but Nehru's motorcade stopped when it reached Pushpa – and Nehru handed her a rose. The story became legend, and people would come from around the city just to see the rose that Nehru had given to her.

My younger brother, Sanjiv, and I attended the finest schools in India and followed our father into medicine – I specialized in endocrinology, Sanjiv in gastroenterology and liver disorders. Upon graduating from the prestigious All-India Institute of Medical Sciences, I followed a well-trodden path and went abroad to finish my medical education. While traditionally young Indian-educated physicians had gone to England to intern, I instead came to the US to intern at Muhlenberg Hospital in Plainfield, New Jersey, in 1970, arriving with $8 in my pocket.

In the US, I established myself as a trusted physician, eventually becoming the Chief of Staff at New England Memorial. At this time I investigated and helped introduce Eastern and Asian health practices to the country. This was the Indian heritage that I brought with me. Growing up, I did not have television in my home, so my impression of the US was formed mostly from books, tales – and Hollywood movies. Among my earliest memories was the day I was on call at the hospital and was asked to 'pronounce' a patient dead. That was not a term used in British medicine. 'Pronounce?' I questioned. 'What does that mean?'

The nurse explained, 'You have to confirm that he's dead.'

'Sister,' I said (in India all nurses are referred to as 'sister'), 'may I have a torch?' A torch being the British word for a flashlight.

The nurse was stunned. 'First of all,' she corrected me, 'I am not your sister. And second, what do you want to do with a torch, cremate him?'

I viewed the US as a land of opportunity. Unlike India, or England, there was no glass ceiling for immigrants or minorities. It was a place where you were judged by your accomplishments, not your heritage, and it was possible to rise by working very hard. In fact, I was surprised to learn that in the US, a medical student could actually challenge the dictates of a professor, which was simply not permitted in the British system.

Initially, I had intended to complete five years of medical training in the US then return to India. But I enjoyed practising medicine there and so decided to stay permanently, which was a difficult decision.

While practising Western medicine, elements of alternative medicine began to ease into my practice. In 1980, I learned Transcendental Meditation (TM), and my physical response to the daily meditation and the questions it raised about consciousness changed my life. I stopped smoking and reduced my consumption of alcohol and coffee. I felt calmer, more energetic and relaxed, and became ten times more efficient in my work. This began what would become my life work, exploring the mind-body connection and bringing its health benefits to those people open to it.

In 1985 Maharishi Mahesh Yogi, who had brought TM to the US, asked me to join him in bringing meditation and the mind-body connection to the West. It was a difficult but remarkable decision for me to make since I had a well-established endocrinology practice. I closed my practice and began working with Maharishi, travelling around the world to bring his message of enlightenment to the public. Meditation, I discovered, enabled people to break down conceptual boundaries and look beyond their normal barriers. It was a way of opening minds. This work brought me into contact with an amazing array of people who were open to personal exploration, including George Harrison and Michael Jackson, who were to become close friends.

While practising allopathy – traditional Western medicine – I began to find that the way we were practising it had too little to do with healing, and too much to do with business. I used to tell people, 'You go to a baker, he's going to sell you bread; you go to a surgeon, he's going to sell you surgery.'

As we were to learn, so much of what is generally considered good medicine is actually of marginal value. Too often I found myself alleviating the symptoms without treating the underlying cause. I also realized that there was so much about the healing process that couldn't be explained by traditional medicine. I remembered reading in a medical school textbook, 'A remission can be attributed to a visitation from God' – but I believed that there had to be another explanation. At that time I began to take an interest in Ayurvedic medicine, a traditional Indian form of alternative or complementary medicine that treats the human being as an entirety rather than a collection of individual bones and organs. It recognizes that the relationship between the mind and the body is essential in preventing and healing illness, and that good health is the result of balance of all parts of life. In 1993, my book, *Ageless Body, Timeless Mind: The Quantum Alternative to Growing Old* was published, and after I appeared on Oprah Winfrey's programme, it sold 100,000 copies in a week and one million copies in six months, spending thirty-four consecutive weeks at No. 1 on the *New York Times* bestsellers' list. As a result, I appeared on the cover of *People* magazine and on other TV shows, which caused millions more people to investigate the possibilities of a greater self-awareness. My series for PBS – *Body, Mind and Soul: The Magic and the Mystery* – drew the highest ratings in public television history and removed any doubt that Americans were open to deeper exploration of their own consciousness. This was the beginning of the mind/body revolution that has now spread worldwide.

In the early 1990s, I met a remarkable man, neurologist Dr David Simon, whose expertise in Ayurvedic medicine, meditation and yoga brought us together to form The Chopra Center for Wellbeing in La Jolla, California. It was the appropriate time to move forward, to leave Maharishi and follow my own path. As David once explained, The Chopra Center was founded on the belief 'that a physician could be more than a technical master of pathology – a doctor could guide his patients to health through his actions, words, and being'. The Chopra Center, currently located in the grounds of La Costa Resort in Carlsbad, California, offers an array of programmes exploring the mind/body connection, as well as medical consultations and instruction in meditation, yoga and Ayurveda.

In the ensuing years I have been honoured numerous times by many organizations, and the mind/body connection has become widely accepted as an essential aspect of good health. I've now written more than seventy books, which have been translated into eighty-five languages. The fame that I have enjoyed has provided me with what I refer to as 'convening power', the ability to bring together people from every conceivable field of research for discussion about the problems we are facing both individually and collectively, in an attempt to create a critical mass of conversation around the solutions. I have been given the extraordinary gift of being able to initiate a worldwide debate about those issues.

In 2002, I founded The Chopra Foundation, which is dedicated to improving health and well-being, cultivating spiritual knowledge, expanding consciousness and promoting world peace to all members of the human family. We offer teaching and resources for health and spirituality for disadvantaged individuals and communities around the world – which even allows me to give back to my home, India, where we sponsor at least three orphanages and have directly impacted the lives of children, as well as host our annual Sages + Scientists Symposium. Each year we award significant grants to three or four individuals for their commitment to science and consciousness. Their exemplary work embodies The Chopra Foundation's own vision – to help create a just, sustainable, healthy and peaceful world.

I have also been able to continue my research, which in many ways has come full circle, as now we are investigating the crossroads of science and consciousness, once again finding ways of breaking through the known barriers. I am working with several world-renowned scientists, trying to bridge that gap between genetics and karma – once again exploring challenging frontiers.

Further Reading & Useful Websites

Ageless Body, Timeless Mind: The Quantum Alternative to Growing Old by Deepak Chopra (2008)

www.chopra.com
www.youtube.com/user/TheChopraWell
www.choprafoundation.org
www.choprafoundation.org/events-initiatives/sages-scientists

TIM FLANNERY

Explorer & Ecologist, Australia

Born in Australia in 1956, Tim Flannery is a renowned scientist, environmentalist, conservationist and writer. He has discovered more than thirty mammal species including new species of tree kangaroos.

Observation of the environmental effects of climate change led Tim to write *The Weather Makers* in 2005, and to help found the Copenhagen Climate Council in 2007. He is now head of Australia's Climate Change Commission.

He has held various academic positions including visiting Professor in Evolutionary and Organismic Biology at Harvard University, Director of the South Australian Museum, Principal Research Scientist at the Australian Museum, Professorial Fellow at the Melbourne Sustainable Society Institute, University of Melbourne, and Panasonic Professor of Environmental Sustainability, Macquarie University. Flannery was the 2007 Australian of the Year. He is currently chief councillor of the Climate Council.

The city of Melbourne was fewer than 120 years old when I came into the world. It was still in many ways a frontier town, and I now realize that its rawness had a big impact on me and my attitude to the environment.

I grew up in the suburb of Sandringham, just 12 kilometres south-east of the city centre on the shores of Port Phillip Bay. It lay at the end of the train line, and in the 1950s and 60s it was the ragged, outer fringe of the city.

My earliest memories are of a natural wonderland. A frog leaping from the back of a toy truck. A hiding place among the ti-trees near our house, with the dappled light falling on the delicate, white, five-petalled ti-tree flowers. A scarlet parrot taking flight from a bush. But by the time I was ten, much of this had vanished.

I remember asking my mother why the last large patch of native ti-tree forest near our home was being bulldozed to make way for a hospital. 'That's progress,' she replied, which had me vowing silently that I'd never have anything to do with progress again.

But there was one last bastion of wildness that progress couldn't touch: the submarine world of the bay. Our house lay just a few hundred metres from the most majestic part of its shoreline – the Red Bluff Cliffs. Sculpted from five-million-year-old sandstone, they stood as a great fluted rampart rising 80 metres above the waves and running for half a kilometre or so along the foreshore before petering out into ti-tree covered slopes. Below it, a rocky peninsula projected out into the bottom of the sea, while just offshore HMVS *Cerberus*, a nineteenth-century ironclad and once the pride of Victoria's navy, lay rusting in the sun, a picturesque breakwater.

While the waters were safe from wholesale alteration, 'progress', unfortunately, could not let even this jewel of a place alone. Around half of the cliff had become a municipal garbage dump, and old cars, refrigerators and other rubbish cascaded down its slope to leak oil into the water below. But this could not blight the life of the sea. If you searched among the rusting rubbish you might find a ribbed murex shell, or the white beauty of an angel's wing clam.

This place was my playground. Standing on the bluff, I could see the schools of fish passing north on their annual migration, and soon my friends and I found secret passages in the bluff, down which we'd carry our fishing lines or masks and snorkels. What a place it was! The seasons of the sea played out there before my eyes. The cloudy waters of summer, filled with plankton, would give way to the breathtaking clarity of winter, when one could see far into the depths. And then, in the spring, the entire place would come to life. Dusky flathead would appear among the rocks in the shallows, their 60-centimetre long bodies so perfectly disguised among the marine growth that often all I could make out were their strange, goat-like eyes peering up at me. And one day the pilchards would arrive, by the tens of thousands, or even millions. To swim out into the migrating school was to be defined by fish, for they would leave a perfect impression of you by swimming around you, just 30 centimetres of clear water separating you and the great fishy accumulation.

Then, when I was ten or eleven, I discovered another bay. Located a few kilometres south of my home, below a forbidding cliff of dark-rust-coloured sediment, lay six- to seven-million-year-old rocks, which had formed on the bed of an ancient, long-vanished bay. They lay exposed on the seabed in just a few metres of water. I remember diving with my mask and snorkel and pulling up a strange, fossilized bone, spoon-shaped and heavy, from among the pink coral and weed.

I took it to the museum, where a venerable curator in a white lab coat proclaimed it the flipper bone of an extinct penguin. Over the years I pulled all manner of fossils from that ancient sea floor, including hundreds of sharks' teeth, some from monsters that would have dwarfed the largest sharks alive, metre-long sections of the jaw-bones of extinct whales, and the spines of stingray's tails and the backbones of seals.

That ancient bay was full of the most extraordinary life. The past, I decided, was a far more interesting place than the present.

Looking back, it seems that at that time I was only truly happy when enclosed in salty water. Then the noise of the world above was cut out, and the sight and smell of my raw, polluted city left behind. When I was twelve my mother and father (who drank too much) parted, and I, the eldest of three children, became the only male in the household. But at school it was all testosterone. St Bede's was a Catholic boys' school in which I served eight years. I will never forget my first day there. The brothers dressed in long, black flowing robes and the boys in blue. It felt like a prison, and I can't recall a single happy day there.

I dreamed of being a scientist like the man I'd shown the penguin bone to, but I did so poorly at maths and languages that I graduated without the requisites to enrol in science at university. So I took on English and history, and wearily resigned myself to being a teacher. But I also took up volunteer work at the museum, where I was taken under the wing of the Curator of Fossils, Dr Tom Rich. He found a few dollars for me here and there, and I enthusiastically undertook any work he suggested.

As I came to the end of my arts degree I realized that I could not face a life of teaching: I desperately wanted to explore the past. Tom suggested that I apply to do a Masters Degree in Geology at Monash University, and I was both astonished and eternally grateful when they accepted me. From there I went on to do a PhD on extinct kangaroos at the University of New South Wales, and from there to work as the Curator of Mammals at the Australian Museum in Sydney. By sheer good luck I had landed the only job in the country that I had really wanted, for it allowed me to study both modern and fossil mammals anywhere in Australasia.

I had long been drawn to the Melanesian island of New Guinea, and soon I was undertaking expeditions there. We would penetrate deep into the jungle, sometimes encountering people who had never seen an outsider, and often discovering species of mammals never described by a scientist. Among the most astonishing of the discoveries were four species of tree-kangaroos, all previously unknown to science. One was the size of a Labrador dog and resembled a miniature panda. It lived under tropical glaciers, high in the alpine meadows and scrubs of Irian Jaya, and had remained unknown to the outside world until 1995. For almost twenty years I roamed the islands of the south-west Pacific, from Fiji to Indonesia – an area almost as wide east to west as Beijing to Paris. In all, the team I led was to discover and name around forty species of mammals new to science.

It was during this work that I first encountered evidence of climate change: everywhere the tree line in New Guinea was rising, and the precious alpine habitats were shrinking. By 1999, I knew what I had to do: give up a life of adventure and become an activist for addressing the threat of climate change. I began researching a book, and in 2005 published *The Weather Makers*, an explanation of climate change and how it might alter our world. It was translated into twenty-three languages and widely read. Then, in 2007, I helped found and then chaired the Copenhagen Climate Council.

For three years in the lead-up to COP15 – the climate summit in Copenhagen – I was close to the international negotiations, assisting business, community groups and the Danish government as they struggled to agree on the best way forward.

Many of us were disappointed, at first, with the outcome of the Copenhagen meeting. It had resulted in a single, brief communiqué called the Copenhagen Accord. But as the months went by, I began to see how important that seemingly slender achievement was. Under it, nations had pledged action to reduce their emissions, and while the action was not sufficient to avoid dangerous climate change, it represented a vital first step. But how was my country, Australia, to live up to its pledge? Australian politics is sharply divided on climate change, and several efforts had already resulted in failure. Then, in 2010, a minority government was formed, with both Labor and Green representation, and an opportunity opened.

In 2011, the government established the Australian Climate Commission, and appointed me as the nation's first Chief Climate Commissioner. The commission was tasked with reaching out to the Australian community (large parts of which remained sceptical about climate science) to explain the need for action. Then in November, the government passed a massive reform bill aimed at addressing climate change.

It established a AUS$23.00 per tonne carbon tax on major polluters, a AUS$10 billion clean energy fund, and a number of other smaller but important initiatives. It also mandated that by 2015, Australia would change its carbon tax into an emissions trading scheme, enabling polluting industries to reduce their emissions by trading carbon emission permits in a market. My country was finally on its way to releasing itself from dependence on the fossil fuels that so threaten our collective future. The fight is far from over – indeed it will go on for decades – but I feel that we have passed a turning point, and from here on in the tide of change will be with us.

Looking back over the fifty-odd years that separate my earliest memories from the present, I can see threads that have influenced my moral philosophy. The brutal destruction of the natural environment I witnessed as a child, for example, has left me with an enduring belief that Nature is precious and vulnerable, and that humans can destroy beautiful things in an almost malicious manner – as if they hate beauty, perhaps because it lies outside their control. I am also wary of the idea of progress. Those in business who promote it are almost always following self-interest. In studying the past, I have been able to enter a world without man, and have found there such an astonishing richness and diversity that it makes our current world look poor indeed. And that impoverishment, I've learned, was all too often wrought by ourselves.

But for all of those influences, without doubt the greatest influence on my thinking has been my experiences in Melanesia, for there I learned a little of what it means to be human, and what possibilities exist for relationships between ourselves and Earth. To walk into an isolated mountain village and make contact, for the first time, with the people living there is to bridge a gap in time spanning 60,000 years. For that's how long ago it was since their ancestors, and mine, went their own separate ways out of Africa. In that time they have become short and dark-skinned, and have developed a sophisticated language and culture, while my people have become tall, pale-skinned and technologically adept.

Such differences are immediately obvious to us all. But what becomes evident only with time is just how much we still share. The nervous smile flashing across the face of a young man, the look of wonder on the face of a child as you open your pack, and the stern wariness of the village big man are all immediately recognizable – the gift of a common humanity that has survived 60,000 years of isolation.

I remember once working in a village where cannibalism was a living memory. I had befriended a young man, and even though we did not understand each other's language, as we hunted in the forest for possums and rats we became quite close. Back in the village he said through an interpreter that he'd like to introduce me to his father. We sat down, and the older man began to tell a long story about how he had got his son. It was on a hunting raid to obtain human meat, he said. The raiding party he was part of had ambushed a hut in which twenty or thirty people slept.

They set fire to the hut and killed people as they emerged. Afterwards, as they were dismembering the bodies, the man had heard a child crying and found an infant hanging from a tree branch in a string bag. He took it home with the butchered bodies of its parents and, he said, he knew immediately that it would grow up to be a good man because it stopped crying as soon as he'd set off with the string bag swinging from his back. His wife was already suckling a baby, and with the protein she got from eating the orphan's parents, he said, she made enough milk to rear both children.

After the old man finished his tale, I was expecting the young man to attack his adoptive father. After all, he had killed and eaten his biological parents. Instead, he held him by the shoulders with clear love. Evidently the story had been told often, and was an affirmation of the bonds that existed between the members of this family.

Such experiences are challenging, but no more challenging than our own cultural practices often are to outsiders. Judging from their reactions, the villagers I stayed with thought that circumcision was the most hilarious practice ever devised; and I will never forget their horrified faces when I used my handkerchief. It was as if I'd placed used toilet paper back in my pocket! Most importantly, I learned that there is almost as much diversity in a village of a few hundred people as exists in all the world, for within its narrow bounds you'll find the quick-witted and the simple, the generous and the selfish, the outgoing and the shy.

Working for twenty years with such people gave me the gift of seeing the world, at least partially, through the eyes of a very different culture. To them, Nature and humanity are one, and thus to damage Nature is to damage ourselves. I feel blessed for having it bestowed upon me. Working in Melanesia also teaches you that life can be short, and that risks are just part of life. So when I faced the choice of staying in my tenured job as museum scientist, or leaving security behind to do something about climate change, I was prepared to take the less secure path.

Climate science has reinforced the lessons I learned in Melanesia all those years ago: the world and us are one. We are, in fact, nothing but animated bits of Earth's crust – a realization that can be both frightening and immensely liberating. And there is no 'away' to throw anything to, and it is only by working together as one human family, regardless of our differences, that we will find a comfortable future. What this has meant for me is that I've been willing to work with business leaders, politicians and community groups from around the world in order to address climate change. And I've found good in all of them. I've even enjoyed discussions with some CEOs about the true meaning of 'progress'!

Cultural differences are of course important, and need to be respected, but ultimately it's our common humanity – and the understanding and empathy we can share as a result – that is our best hope for our survival.

For the past year I've been working in a role with a political edge to it. Me being Australia's Chief Climate Commissioner has irritated some, in part because the conservative side of Australian politics includes many climate sceptics as well as many who are doing very well out of the status quo. Regrettably, their numbers also include those in the media who are willing to stoop to publishing malicious libel to discredit people such as myself, as well as others who are happy to issue death threats to those who warn of the dangers of climate change. During one town hall meeting that the commission conducted in Western Australia, I was approached by an elderly couple who said that they didn't give a damn about the future. They were happy to inflict environmental damage on future generations (they had no children) so long as they could live out their remaining years in comfort. I'm still trying to come to terms with this, and must admit I've not yet found a way to reach out to them. Perhaps it's only those around them – the members of their community – who can influence them.

I'm very proud of the way my country has taken action on climate change. It was hard to achieve, and regrettably it's fractured our society, creating fear that the economy will be damaged as a consequence, or that the UN is trying to control our lives. Now, I think, it's the time for peacemaking. For reaching out to the common morality of care of family and environment that most of us share, regardless of differences in religion, politics or wealth. For in it lies the only chance of survival.

Further Reading & Useful Websites

The Weather Makers: The History and Future Impact of Climate Change by Tim Flannery (2005)

Here on Earth: A New Beginning by Tim Flannery (2010)

www.theweathermakers.org
www.timflannery.com.au

JANE GOODALL

Scientist & Conservationist, UK

Born in London in 1934, Dame Jane Goodall is an anthropologist, ethologist, conservationist and the recipient of multiple awards including the UN Messenger of Peace. She is best known for her forty-five-year study of social and family interactions of chimpanzees in Gombe Stream National Park, Tanzania, and she still works extensively on conservation and animal welfare issues. In 1977, Jane founded the Jane Goodall Institute, which, in addition to conservation, promotes sustainable livelihoods in local communities. The Jane Goodall Institute has created a worldwide network of young people who are committed to taking care of their human community as well as the animals and their environment. Core values include a recognition that all living things – people, animals and the environment – are interconnected; that every individual has the ability to make a positive difference; and that flexibility and open-mindedness are essential to enable us to respond to a changing world.

There are four individuals who have inspired me in my activism and in my life. The first one is my amazing mother. She supported all my childhood passion for animals. Lots of little girls – and little boys too – love animals, but they don't always have such an understanding mother as I had. When my mother found I'd taken earthworms to bed with me, she didn't get angry and throw them out. She just said quietly that they would die without the earth, so I helped her to take them back to the garden.

My first vivid memory is when I was four-and-a-half years old and we went on a holiday to the country. Here I was, this animal-loving little girl from the city, where there are cats, dogs, pigeons and sparrows and not much else; now I was meeting cows, pigs, horses, sheep, face to face, all grazing out in the fields.

One of my jobs was to help collect the hens' eggs. I was getting them out of the nest boxes and putting them in my little basket, wondering where on a hen was there a hole big enough for that egg to come out? I looked and looked and couldn't see one. I was asking everybody, 'Where does the egg come out?' But no one answered me to my satisfaction, so I decided I would have to find out for myself.

I saw a hen climbing up into the hen-house where the nest boxes were and I thought, 'Ah, she's going to lay an egg.' And I followed her. Well, that was a mistake. She flew out, probably terrified. So, that wasn't a good place to wait. I then went into an empty hen-house and waited, and finally as dusk was falling my mother sees this excited little girl rushing towards the house all covered in straw. And instead of being angry with me for making them all worried because no one knew where I was, she saw my shining eyes and sat down to hear the wonderful story of how a hen lays an egg.

So, there you have the making of a little scientist: curiosity, asking questions, if you don't find the answer straight away, you think you will find out for yourself, you make a mistake, you don't give up and you learn patience. And my mother was helping the development of this little scientist.

Then I found the books about Tarzan of the Apes. And, of course, I fell in love with this glorious lord of the jungle. And what did he do? He married that other Jane. I was really jealous and I was sure I would have been a better mate for Tarzan!

So, when I was eleven years old I decided that when I grew up, I would go to Africa, live with animals and write books about them. Not surprisingly, everybody laughed at me. It was the Second World War period and Africa was still thought of as the 'dark continent', very far away, no 747s going back and forth with tourists. We didn't know much about this dark continent. There were rumours of poisoned arrows and cannibals – but also, all those wonderful animals. The biggest problem of all for me was that I was the wrong sex. Back then, girls didn't do that sort of thing. But the one person who never laughed at me was my mother. She would say, 'Jane, if you really want something and you work hard and take advantage of opportunity and never give up, you will find a way to do what you want to do.'

When I left school my friends went to university, but we couldn't afford it. And in those days you couldn't get a scholarship unless you were good in a foreign language, and I wasn't. It was my mother who said, 'Well, do a secretarial course, and then maybe you can get a job in Africa. That'll take you closer to what you want to do.' So, I did. I was working in London when I got a letter from a school friend, inviting me to Kenya for a holiday. Yes: opportunity, but still no money. So I gave up my job and went home where I could live for free, worked in a hotel close by and saved up. Eventually I had saved enough money for my return fare to Africa by boat.

I was twenty-three when, in 1957, I waved goodbye to my family, my friends and my country, and set off on this amazing adventure. For me, every day is an adventure because we never know what we're going to learn, who we're going to meet or what opportunities will come our way. But that time, it was a real adventure: setting off on my own to go and stay with a friend.

Soon after I arrived, I got a job in Nairobi. There I heard about Louis Leakey. He was another powerful inspiration. I went to see him at the Natural History Museum and he asked me all kinds of questions about the animals there. Because I had followed my mother's advice, and I had continued to read books about Africa and animals, and because I had spent hours in the Natural History Museum in London, I was able to answer many of his questions. I was ready for this opportunity. He gave me a job working as his assistant. And he let me go with him to a place that's now very famous: Olduvai Gorge on the Serengeti Plains.

In those days, Olduvai Gorge wasn't famous at all because no human fossils had been found there, only the fossilized remains of various animals. And all the animals were there back then: the giraffe and the zebra, the rhinos and the lions. I was allowed out on the plains after a hard day of digging for fossils.

Sometimes we met a rhino. Once we met a young male lion; he was fully grown with little wisps of hair on his shoulders – and he was curious. He had never seen anything like me before. And although it was a bit scary, it was very exciting.

I think that's when Louis Leakey decided he would offer

me the opportunity to go to Gombe in Tanganyika (western Tanzania today) and learn not about any old animal, but about the animal more like us than any other. Of course, I said yes. But it wasn't easy. It took a year before I could start on this amazing experience because, first of all, who was going to give money?

I had no university degree; I had just come straight from secondary school. I was English. And I was a girl. But eventually a wealthy American businessman said to Louis Leakey, 'OK, here's money for six months, we'll see how this young lady does.'

The second problem: The Tanganyikan authorities refused to take responsibility for a young girl arriving on her own. But in the end they said, 'Oh, all right – if she has a companion.' So, who volunteered to come? That same amazing mother. She packed up in England; she could come for four months. It was a shoestring of an expedition. We had one ex-army tent that leaked: no fancy sewn-in-ground sheets, but just a piece of canvas on the ground and the sides of the tent rolled up to let the air in – but this also let in the spiders, the scorpions and the snakes, none of which my mother liked very much. But she never complained. And we had some tin cans to eat from, and a tin plate and a tin cup, and that was about it. I mean, how amazing? How many mothers would do that?

She did two things for me. One: In those days when I first began, the chimpanzees – who were very conservative, who had never seen a white ape before – took one look at me and vanished into the undergrowth. I would get back depressed because I knew if I didn't see something exciting before the money ran out, that would be the end: no more study, I would have let Louis Leakey down. But there was Mum in the evening – and I never got back until it was just about dark. We would share our simple supper and she would console me and point out what I had learned. She'd point out that I was learning the foods that chimpanzees were eating, that I saw how they made a platform or nest of branches up in the trees each night. I was developing an understanding of their social structure, where they move around sometimes in small groups, sometimes singly, sometimes small groups meeting up into a large gathering – continual coming and going within this community. Thus she really boosted my morale and released me from my disappointment of not actually being with chimpanzees.

Two: She started a clinic. She wasn't a doctor or a nurse, but she cared about people, and her brother was a doctor. So, we had all these simple medications like aspirins and bandages, Epsom salts and saline drips. She made some amazing cures. She became known as a white witch doctor, because she practised white medicine. Thus she established for me an amazing relationship with all the local people, and that has stood me in good stead ever since.

The third of the individuals who inspired me to be an animal activist was a chimpanzee called David Greybeard. It was he who featured in my breakthrough observation. I can never forget that day in 1960, about five months into the study, when I was walking back through the forest. It had been raining, I was wet and cold, and I suddenly saw a dark shape huddled over a termite mound. I stopped and I peered with my binoculars. I saw a hand reach out and pick a piece of grass. And I could see a chimpanzee was using this as a tool – pushing it down into the termite mound, into the passage of the termites. Waiting for a moment, pulling it out, picking the termites off with his lips. And then I saw him pick a leafy twig and strip the leaves.

Why was this so exciting? This chimpanzee – the one I'd named David Greybeard for his white beard – was using a tool. He was making a tool by modifying an object. And this was exciting because at that time, it was thought that using and making tools differentiated us, the humans, more than anything else from the rest of the animal kingdom. We were known as 'man, the tool maker'. And when I sent a telegram to Louis Leakey, the reply came back: 'Now we must redefine man, redefine tool, or accept chimpanzees as humans.' That observation of David Greybeard using tools was a real red-letter day.

So Greybeard was the first chimpanzee I came to know. Not only did he demonstrate tool-using and the fact that chimpanzees hunt for prey and share the carcass, he was also the first to begin to lose his fear of me. When I appeared in a group, he sat there calmly looking at me; the others with big wide eyes saw him looking at me and said, 'Well, she's not so frightening after all.' So, really, Greybeard helped me to move through a doorway into a magic world, the world of the wild chimpanzees. And wasn't I lucky to be the first person to really explore that world in depth?

Looking back now over more than fifty years from when I first began in Gombe National Park in Tanzania – that's amazing. That's the longest unbroken study of any group of wild animals in the world, and I am still there. And I am still learning.

Looking over this half century of knowledge and exploration, what do I find that's really fascinating? There is so much to be fascinated by but, above all, I am amazed to know how chimpanzees are very much like humans.

Biologically, the DNA of chimps and humans differs by only just over 1 per cent. The blood of a chimpanzee is so like ours that you could have a blood transfusion if you matched the blood group. You could not take blood from a gorilla. Chimpanzees are biologically more like us than gorillas. The immune system is so similar that they can catch or be infected with all known human contagious diseases.

Another fascinating fact relates to the brain. The brain of a chimpanzee is so much like ours. The main difference is in its size. It shouldn't surprise us, then, that they are capable of intellectual abilities that we used to think unique to us. One by one, many of those attributes that were supposed to mark humans as separate and unique have been broken down through observations of chimpanzees and other animals.

Chimpanzees have a long childhood. The mother has her first baby when she's twelve or thirteen. She then has one child every five or six years. That's a long period during which the child is riding on the mother's back, sleeping with her at night and suckling. This long childhood is as important for chimpanzees as it is for humans because they, like us, have a lot to learn.

They learn things like tool-using behaviour. There are eight different ways chimpanzees use objects as tools at Gombe alone. In every place in Africa where chimpanzees are being studied, they use different objects for different purposes. Chimps are like humans because the young ones learn by observing the adults and imitating them and practising. That's one definition of human culture: behaviour that's passed from one generation to the next through observational learning. So, we can say that chimpanzees have their own primitive cultures.

We can discover the extent of chimpanzee intelligence in captive situations, where chimpanzees can be rewarded. It's like sending children to school and encouraging them to learn. Chimpanzees can learn American Sign Language (ASL). They can learn about 400 of the signs of ASL, and they can use them to communicate with each other – although they prefer to use their own postures and gestures.

It was fascinating for me to watch the development of the relationship between the mother and her offspring, to see how during these five years, when the child is totally dependent for transport, food and learning, the bonds get stronger and stronger. And then when a new baby is born, the older child doesn't leave; it remains with the mother, and so the bonds get even stronger. And the bonds develop between brothers and sisters too.

It was a shock to me when I first realized that chimpanzees, like us, had a dark side to their nature; in interactions between neighbouring groups and communities in particular, there can be violent behaviour. Groups of males patrolling the boundary of their territory may give chase if they see strangers from a neighbouring group, and they may attack, leaving victims to die of wounds inflicted. But we can take comfort from the fact that they also show love and compassion. They can show true altruism.

Imagine you're with me and we're walking through a forest. It's lovely and dim and green under the canopy, with little specks of sunlight dancing down. And we're following a male chimp called Satan. He's not very wicked, but when he was a kid he stole a manuscript from my mother, who'd come on a visit. We had to bribe him with a banana to get it back. But that was long ago, and now, here he is: twenty-three years old, in his prime. And suddenly he hears sounds of an excited group of chimpanzees feeding in the trees ahead. Imagine about twenty chimps making sounds. So, now Satan is all excited, his hair bristles, he hurries along the trail.

He comes to this big tree filled with ripe fruit and feeding chimpanzees, and he rushes up. There's a bunch of red figs: delicious. He goes straight there. Well, there's already a younger male, about two years younger than Satan, feeding there. But Satan is dominant, threatens that younger male away and begins to feed. The young male starts screaming. There's a special call they give. It's a call for help. And unknown to Satan, that young male's older brother is feeding higher up in the tree. And now, hearing his kid brother in trouble, he comes swinging down. The two brothers attack Satan together. Now Satan screams. To my amazement, a very old female – whose teeth are worn to the gums, who's shrivelled with age, who weighs about half of each of these battling males, who's probably closer to sixty years old, and who has been feeding quietly in the canopy – comes swinging down. She drops her frail self onto these three males, and with her little fists starts hitting at the two brothers. They are so surprised, and mildly threaten her – while Satan gets away.

And that is Satan's old mother, Sprout.

When a mother dies, her older child may care for an infant. If the infant is over three years old, there is a good chance of survival if there's an older brother or sister to look after that child. One such infant was Mel. He was three-and-a-half years old when his mother died, and he had no older brother or sister of his own; he was alone in the world. But to our amazement, a twelve-year-old adolescent male named Spindle adopted him – waited for him in travel, let little Mel ride on his back. If Mel begged for food, Spindle would share; if Mel crept up when Spindle made his nest at night, Spindle would draw him close and they would sleep curled up together. He would even rescue little Mel if he got too close to some aroused males competing for dominance. If an infant gets in the way as they charge across the ground, they may actually pick it up and throw it. It seems they lose their inhibitions. A mother's job is to keep the child away; Spindle took on that job as well, and saved Mel's life.

So there they are, these chimpanzees with their similarities to humans and their intellectual abilities so like ours, their postures and gestures: kissing, embracing, holding hands, patting one another on the back, shaking a fist, laughing, tickling each other. They do these things in the same kind of way as we do.

These amazing beings have really and truly helped me to blur the line once thought to be so sharp, dividing us from the rest of the animal kingdom. They force us to admit that we humans are not the only beings on this planet with personalities, minds and feelings. This gives us cause to show new respect not only for the chimpanzees, but also for the other animals with which we share our planet. Back in 1960, when I began, it was very different. I was told I'd done everything wrong: I should have numbered the chimps, and naming them wasn't scientific; I couldn't talk about chimpanzees having personalities or minds or feelings, because those were unique to humans.

But fortunately, I thought back to a teacher I had as a child, a teacher who taught me that animals absolutely do have personalities and minds and feelings. And that teacher was my fourth individual, my dog, Rusty – who, like my mother and Louis Leakey and David Greybeard, inspired me to become an activist. What I learned from him helped me to have the courage of my convictions in spite of those erudite professors who didn't like the way I did things.

So, it's a bit sad to find that chimpanzees, who really have been like ambassadors from the animal kingdom to us, are vanishing in the wild. They're becoming extinct, like so many other animals around the world today. They're disappearing because of the destruction of their habitat and ever-growing human populations. They're disappearing because they are being hunted for food – not to feed hungry people, but because of the commercial hunting of wild animals, which is facilitated by the intrusion of new roads created by logging companies. The bushmeat trade supplies all kinds of wildlife – smoked, cut up and offered for sale in markets to the rich who have a cultural preference for it. It is often exported to Europe and the US.

Even the chimpanzees in Gombe are endangered. When I flew over the area in the late 1980s and looked down from a small plane, I was absolutely shocked to see the habitat outside the park – it's tiny, just 30 square miles of beautiful forest with crystal clear streams running down to the water of Lake Tanganyika. All the trees around the park had gone. And this is steep, hilly country, so as the trees went, there was terrible soil erosion, and the thin layer of topsoil was being washed down into the lake. There were more people living on this landscape than it could support.

The question that came into my mind as I flew over was, how can we even try to protect these famous chimpanzees if the people living around them are struggling to survive? And that's what led me to work in this area to improve the lives of the people living in the villages around the Gombe Stream National Park. We started by employing local Tanzanians to go into the villages, listen to the elders and ask them what they felt would improve their lives, what would make things better for them.

The villagers were concerned about health and the education of their children. So we began to work in these programmes. Then we gradually introduced the other elements of the programme: farming methods most suitable for the steep, eroded land; ways of reclaiming overused farmland so that within two years it can be productive again; working with groups of women, giving them opportunities to take out tiny loans. We also provide scholarships for girls to stay in school beyond the primary level. We provide information about family planning. We work with women because all around the world it's been shown that as women's education improves, family size drops.

And we work to ensure that the projects that result from these tiny loans from the micro-credit banks are environmentally sustainable.

One of the things we've done in these villages is to introduce our global youth programme, Roots & Shoots. This began in Tanzania in 1991, and is now in more than 130 countries around the world, involving thousands of empowered young people in projects to benefit their community, animals and the environment we all share. There are members from pre-school, first grade, second grade and right on up to and past university, and we're in refugee camps and prisons and retirement homes and so forth.

So how did this begin? Well, it began because when I realized what was going on with chimpanzees, how they were vanishing right across Africa, I realized I had to stop living this lovely dream life: working in the forest, learning about these amazing beings, doing some research, writing and lecturing. I had to stop and go out and talk about what was going on and raise awareness. And as I was travelling around in Africa, I began to find out what was actually going on there. I began to have a sense of the degradation of the environment and the problems faced by the people: the deforestation leading to soil erosion, leading to desertification, leading to loss of biodiversity, leading to huge problems for the people living there, who began to live increasingly in a vicious cycle of poverty and over-population, hunger and disease.

I began to realize that the environment is being destroyed in two ways. One: by very poor people who destroy the environment simply because they have to find some way to feed their families. They are forced to cut down trees to grow crops, or to keep goats, which destroy the environment, because they have no alternatives. They know perfectly well that because of their activities, they're going to change what may start off as lush forest into a desert, but what else can they do? And then two: you have the unsustainable lifestyles of the elite communities who take far more than their fair share of non-renewable natural resources. So, I decided that I must start travelling outside Africa, going around in Europe and the US and in Asia, and speaking about all the different ways in which we are harming the planet.

We're poisoning the air, the water and the land. We have children being born in environments where the air they breathe, the water they drink, the food they eat is actually making them sick. We are burning up fossil fuel in a greedy way, adding to the greenhouse gases that are leading to climate change.

Water is a huge problem. The surface water is shrinking, the water tables are dropping, the great aquifers are becoming endangered. The pollution is washing down with the rain into the streams, the rivers, the lakes and the sea. The fish that we eat are becoming contaminated as well as being overfished. We're eating more and more meat, which means that animals are being farmed not only in an inhumane way, but in intensive factory farms where to keep them alive they must routinely be fed antibiotics. The bacteria are building up resistance and so superbugs are being created. The intensive farms are the second-biggest polluters and creators of greenhouse gases on the planet. And as more people around the planet get more money and eat more meat, what is going to happen?

The American biologist E.O. Wilson has commented that if the entire population of the planet today were to attain the standard of living of the average American or European, we would need three new planets. But we don't even have one new planet; we only have this one.

So, what's gone wrong? How is it that we're destroying our only home?

Do you think it's because we've lost something called wisdom? The wisdom of the indigenous people who would gather together to make a major decision and ask how the decision they make today will affect people seven generations ahead? And how do we make decisions today? On how this will affect me or my family now? How this will affect the next shareholders' meeting, three months ahead? How this will affect my next political campaign? These are the kinds of criteria we're using to base our decisions on today.

Is there a disconnect between this very clever brain of ours and the human heart, the seat of love and compassion? If we don't have a grounding in this very humane part of us, we create a very dangerous animal indeed, the animal that can go out and make weapons of mass destruction and kill others far away by pressing a button, and destroying the environment to the detriment of the children of the future.

As I began travelling around the world, I met many young people and so many of them seemed to have lost hope for the future. They were depressed, they were apathetic, some of them were angry. When I talked to them, they more or less said the same thing: 'We feel this way because we feel you've compromised our future and there's nothing we can do about it.' We have compromised the future of our young people today. I've got three little grandchildren. I think how we've harmed the planet since I was their age, and I feel a kind of desperation.

But it's not right to think that there's nothing we can do about it. There is a lot we can do about it. This is what led to the Roots & Shoots programme. Imagine an acorn – little roots come out, a little shoot comes out – and how small that looks. But there is so much energy, so much life-force in that seed, that those little roots can push their way through boulders to reach the water, and that little shoot can work its way through crevices in a brick wall to reach the sun, until the boulders and the brick walls – all the problems we've inflicted on the planet – are dispersed.

The most important message: every one of us makes a difference every day. And if we would just spend a little bit of time thinking about the consequences of the choices we make each day – what we buy, what we wear, what we eat – there is so much we can do. Collectively, that will start to make bigger changes as more people understand that their own life does make a difference. I meet so many people who think about the problems of the world and feel helpless, feel there's nothing they can do as one person, and so do nothing. But as people begin to work together – making these small, correct decisions, thinking about what they can do to reduce their own ecological footprint – we start to see the changes we must have. And kids get this. They really do.

What do the Roots & Shoots groups do? They decide for themselves. The programme is youth-driven. And even quite small children have a good idea of what they want to do. So, in a group, they will sit around and talk about the local problems, decide which ones they care about, and choose projects they can take action on. And one kind of project will help animals, including domestic animals; one kind of project will help people; and one kind of project will help the environment. Running throughout is a theme of learning to live in peace and harmony and to break down the walls that we build between people of different nations, different cultures, different religions, and between us and the natural world. Kofi Annan made me a Messenger of Peace because I explained Roots & Shoots to him and said: 'Wherever we go, we are sowing seeds of global peace.'

Everywhere I go, people ask me, 'Dr Jane, do you really have hope for the future? You've written a book called Reason for Hope. You give talks called "*Reason for Hope*". You've seen the forests disappearing in Africa, you've seen the chimpanzees decrease in numbers, and you've seen terrible examples of inhumanity. Do you really have hope for the future?'

Well, my four reasons for hope are very simple, probably very naive, but they work for me. The first one is these young people: the Roots & Shoots groups and other similar youth around the planet. Some of the Roots & Shoots groups are doing extraordinary things, and sometimes it requires a lot of courage. These young people are my source of hope.

Then there are groups of people who work for three, four, five, six years on a nature restoration project, like removing the exotic plants from wetlands or from a piece of prairie, to restore the environment to how it once was. These groups are my great reasons for hope: the determination, the enthusiasm, the energy, the commitment and the courage of people all around the world when they know what the problems are and they are empowered to take action. Where does my energy come from?

It comes from these amazing young people.

My next reason for hope is the resilience of Nature. I witnessed the trees at Gombe that grow out of seemingly dead tree stumps. I was in the redwoods and saw how these trees won't die. Even if you cut down a giant tree and you leave the stump, in time the roots put up new saplings and eventually they will make a circle around the stump and join together to make another giant tree. They won't die.

They will restore themselves.

And animals on the brink of extinction can be given a second chance. Think of the California condor. There were twelve of them left at one time, and through the passionate dedication of some biologists who felt they could not allow this amazing creature to vanish, now there are more than 200 condors flying in the skies above four states of the West Coast of the US.

Another reason for hope is the indomitable human spirit. And everywhere, in all walks of life, there are people tackling seemingly impossible things. They can be people like Nelson Mandela, who spent twenty-seven years in prison, with the amazing ability to forgive, so that he led his nation out of the evil regime of apartheid without a blood bath; people like Muhammad Yunus, who designed the Grameen Bank when all the other big banks told him it was impossible, and has now done probably more to help the poorest of the poor than anybody else. And then there are people living all around us, who we may just pass in the street and never know we've walked past a person who's had to overcome seemingly impossible odds: physical disabilities; social problems; fleeing from a country, arriving with no money, no word of English, no friends, and somehow making a life for themselves and their families.

Then think about the human brain. The human brain attached to the human heart is indeed an extraordinary mechanism. And we are now beginning to create technology that will help business to be more environmentally friendly. We're using our brains to work out things like carbon trading, carbon credits. We're using our brains to work out how each one of us can walk through life with a smaller ecological footprint. All of these things together give me hope for the future.

There is an enormous amount of hope, which lies with each one of us. We all have to do our bit. It isn't just going to happen. Everywhere in our world there are problems.

But wherever there are problems I find a group of caring, compassionate, dedicated, courageous people who are working for little or no money, who are risking their health, sometimes risking their lives, to try to put those problems right. And it's this that gives me the greatest hope.

Further Reading & Useful Websites

A Spiritual Journey by Jane Goodall (2021)

Reason for Hope by Jane Goodall (2004)

www.rootsandshoots.org
www.janegoodall.org

ROGER HALLAM

Environnmental Activist, UK

Born in 1966, Roger Hallam is a Welsh farmer and co-founder of Extinction Rebellion and Insulate Britain. He studied for a PhD at King's College in London researching how to achieve social change through civil disobedience and radical mobilization before launching Extinction Rebellion with the "Declaration of Rebellion" on 31 October 2018 outside the UK Parliament. Extinction Rebellion is now a global environmental movement. Roger Hallam has been arrested multiple times, been on hunger strike twice and been imprisoned three times in the last three years. In 2019 he published *Common Sense for the 21st Century: Only Nonviolent Rebellion Can Now Stop Climate Breakdown and Social Collapse* and the essay entitled 'The Civil Resistance Model' in *This Is Not a Drill: An Extinction Rebellion Handbook*.

"THE PRICE OF LIFE IS DEATH"[1]
GEORGE MALLORY

I spent twenty years growing vegetables on a small Welsh farm. The highlight of my week was going to buy hummus on Saturday mornings in the local town. For years I hardly saw anyone and I didn't much care. I was happy enough to leave the world to its ways and get on with weeding my carrots. But the world's ways, as billions of farmers know, are now destroying those carrots, because farming each year now involves a trip to the casino. You shake a dice on extreme heat or cold, extreme rainfall or drought. There's no fun in betting on the farm every year. That's why I am in rebellion. Because my livelihood was destroyed and soon everyone's livelihoods will be destroyed. And nothing can happen without food in your stomach as the world will soon discover.

The climate is breaking down. There is no hope of this stopping within the existing system. It was out of this burning of our Enlightenment egos that Extinction Rebellion (XR) was born. It was the transformation of existential rage into organizational discipline – an unlikely alchemy. This was before XR got swamped with the niceness of middle-class hope. I spoke recently to a bunch of XR activists on Zoom in Australia. I said Australia is done. Because it is. Within twenty years, the 20 per cent burning down of Australia's forest, the 80 per cent of Australians with the health effects, the almost three billion animals killed or harmed in the 2019–2020 bushfires will be a normal summer. In forty years, it will be a cool summer. It's all there in the peer review science papers.

---

1 *Into the Silence: The Great War, Mallory and the Conquest of Everest* by Wade Davis (2011)

Sydney will be evacuated within a decade. True, it could be two decades, but then it could be next year. The faces on Zoom looked at me like I had insulted their god – which of course I had – the god of progress, the god of hope.

In California, where I also facilitate XR online meetings, they had the biggest ever forest fire in 2018 – two million acres – 2 per cent of the state burnt down. In 2020, the fires doubled in size, 4 per cent burnt down – 4 per cent of the state. Do your GCSE maths. By 2030, up to 30 per cent of the state will be burning down each year. Los Angeles will be evacuated.

Hints of the apocalypse to come are everywhere in trade magazines and regional press. This week we read that the dams on the Colorado River will be dry by the mid-2020s. 40 per cent of Russian buildings in the Arctic Circle are subsiding due to the melting permafrost (it's six degrees Celsius above pre-industrial levels up there). A Russian ministry official predicts 30 per cent of the federal budget will be committed to propping them up by 2050. Add in the compounding effects of the collapse of the global trading system, the routine failure of the Russian wheat harvest, and the migration of hundreds of millions from the wet bulb zones of the Middle East, and any self-respecting political scientist is going to tick the box 'state collapse' within a generation. And that's just Russia.

We are being systematically lied to. Three climate scientists recently went on record to say that they knew of no scientists at the 2015 Paris conference who thought that staying under 1.5 degrees Celsius was feasible. That was six years ago. Sir David King, the former Chief Scientific Advisor to the UK government, said in 2021: 'We have to move quickly. What we do in the next three to four years will determine the future of humanity'. That's Oxbridge speak for billions of people will be slaughtered or starved to death. What else do you think he means? It is time to grow up and see the world as it is. Unless the information bypasses your head and hits you in the stomach, you are not truly facing the prospect of annihilation.

Why isn't the climate movement succeeding in reducing carbon emissions? While I was doing research at King's College, a few students and I spray-painted around £10,000 of damage to the central university hall to get the college to divest from fossil fuels. The vice principal was there within five minutes. He said, 'Roger, this sort of thing shuts down the conversation.' It was the first conversation I had ever had with him. When the carbon regime is threatened, it always confuses moral disapproval with political effectiveness. After being arrested I started a two-week hunger strike. Within those fourteen days, the college turned around its whole policy on the climate and agreed to 100 per cent divestment. It was the quickest divestment campaign in the world. Did environment campaigners set up a programme to promote criminal damage and hunger strikes to create wholesale climate policy change in Western institutions? Nope. Not even a flicker of interest. The climate movement reformist culture trumps considerations of objective effectiveness. Why else is it rejecting the moral no-brainer of mass civil resistance against governments engaged in the greatest ever genocide of life on this planet?

Extinction Rebellion was set up do what it says on the tin – to organize a rebellion against the British government. It was not supposed to be a brand, like calling a restaurant chain 'Revolution'. In my wildest dreams, I thought we would get 3,000 sign-ups and that, by now, half of them would be serving prison sentences. Instead, we got 200,000 sign-ups and one person is in prison. The April Rebellion of 2019 changed British history: 10,000 people in London for ten days occupying five sites with 1,200 arrests while maintaining non-violent discipline. The back story is that it almost didn't happen. There was Brexit; people doubted we would get the numbers; there was fear that people wouldn't want to get arrested, or that if they did we would put off the public. The small number of people who pushed it through argued that we cannot afford to take the risk of not taking the risk. We are out of time.

April 2019 changed the conversation in the country, got Parliament to declare a climate emergency and 50,000 people joined in a month. XR was named the most influential global actor on climate in 2019 and it was set up in seventy countries. But all this didn't actually mean anything. No actual carbon emissions were reduced because it didn't push through to material victory. The plan was to come back in four weeks after April and finish the job. Instead, people had to go on 'regen break'. Some of us proposed we close down Heathrow Airport with toy drones flown at head height. Panic broke out and it failed to get movement support. A few carried on and I went to prison for five weeks. I looked over my shoulder, but there was no army following. The momentum had disappeared into a mere mailing list. However long that list is, it means nothing unless it translates into a critical mass of disruption. And it didn't, because climate movements don't do what is needed to win.

The deeper reason for this desperate state of affairs is that today's climate movement holds us in deep fear and denial of death. In the 1920s, some survivors, including George Mallory, of the bloodlands of the First World War – the fields of mud and splattered bodies – climbed Everest but did not return. They were the antithesis of today's climate movement. They saw death and death educated them on what real life is: they risked death in order to be truly alive. It's a life philosophy completely at odds with the 'duty-of-care' saturated 'compassion' that obsesses today's gatekeepers of the climate movement and prevents people rising into revolt. It's my contention that only this philosophy will save the human race because it provides the vital basis for the courage to enter into resistance.

We have, according to Sir David King, three to four years to get this emergency decarbonization show on the road. This requires civil disobedience resulting in several thousand arrests and/or several hundred people in prison to force legislative change in an average Western country. It's not complicated – there is always a tipping point and this is broadly it. I am not against the 101 other climate activism bits and pieces people do. I am not against all the vital specialisms that are needed to bring in decarbonization – the academic researchers, the engineers, skilled manual workers. But nothing will happen until a critical mass of people pull down the walls of the climate regimes through civil resistance. Only when this happens will the social forces be released to create the great transformations that we all know we need.

So what's the plan? The plan is youth. When I was young I dropped out of university, got arrested every other weekend, and went to prison three times before I was twenty-one. And I loved every minute of my resistance to the threat of nuclear holocaust. This is the archetypal role of youth in human history – the exuberant drive into the ecstatic chaos of life. The laughing in the face of death. The 'come and do your worst' spirit of rebellion.

We have been here many times before. In 1962, polite, smart tie-wearing young people set up the Students For a Democratic Society and produced a nice bunch of words called the Port Huron statement. Seven years later, Abbie Hoffman would be doing handstands in front of a judge during the Vietnam protests. Everything changed in a matter of years.

During this decade, Black civil rights organizations ran a number of dramatically successful campaigns. One of the most famous was the Birmingham, Alabama, campaign of 1963. After failing to make an impression by Martin Luther King going to jail and the difficulties of mobilizing the adult Black residents of the city, the idea was hatched to involve the city's children and young people in an ongoing escalation of mass participation civil disobedience. This involved thousands of pupils and students leaving school to illegally march through the centre of the city. Word was spread via local radio stations and meetings, which promoted the methods of civil disobedience. A 'D-Day' was set when the mass action would begin. The authorities opted for a repression response, arresting 1,000 of the protestors on the first day and 3,000 on the second day. This triggered the backfiring effect. Thousands more children left their classes to protest and fill the jails. 'The fear had gone' as one officer reported. After a week, there was no end to the mass protest. The authorities had lost control and the opposition collapsed. Decades of segregation policies were overturned in a week. This is the power of mass civil disobedience.

The same thing happened in the Leipzig Monday demonstrations in East Germany in 1989. A pastor was extremely disillusioned with the communist regime and lead his community to a public demonstration over a weekend. Approximately 6,500 people attended. Initially, the local security forces did not interfere with a small Christian demonstration. Encouraged by this success and the police reaction, the following Monday 17,000 people came out on the street. Unsure what to do, the authorities contacted their superiors in Berlin. A message came down the line to shoot at the demonstrators. However, the next Monday there were 60,000 people were on the streets and the police could not bring themselves to shoot that many people. They disobeyed their orders. The following weekend 105,000 people turned out. The fear had gone, and this is the moment when the tide turned and the regime collapsed shortly afterwards.

Passivity is not due to apathy, but to the anti-life culture of the present climate movement that's imposed upon us. But it's breaking down. Since Western countries have started to emerge from the Covid-19 pandemic, there have finally been the initial moves towards classical civil resistance. A few of us started Insulate Britain in January 2021. As I write this, people have been making headlines for blocking motorways week after week. All around the world young people are getting with the programme: at first as single individuals and now in small groups. In 2022, they will start bursting into the political space and it will not be long before they, too, are doing handstands in the courts. This plunging into life will animate a new revolutionary climate rebellion. People do not just lie down and wait to die – least of all youth.

We can no longer afford the luxury or indulgence of hoping for perfect political and economic conditions. Ignoring the social science on radical political change is as immoral and criminal as ignoring the climate science. There are no easy options anymore. The time of playing the nice guy is gone. There is only one action that leads to true self-respect and that is the act of rebellion. We need to fight the good fight and, when our time is done, be able to say, 'Yes, I played my part.'

Let's get to it.

Further Readings & Useful Websites

Common Sense for the 21st Century: Only Nonviolent Rebellion Can Now Stop Climate Breakdown and Social Collapse by Roger Hallam (2019)

This Is Not a Drill: An Extinction Rebellion Handbook including Roger Hallam's essay 'The Civil Resistance Model', pp. 99–105 (2019)

POLLY HIGGINS

International Environment Lawyer, UK

Born in Scotland in 1968, Polly Higgins was a barrister, author and environmental activist, voted one of the 'World's Top 10 Visionary Thinkers' by *The Ecologist*. Polly's campaign to make ecocide a crime subject to trial by jury attracted worldwide attention in 2011, when a mock trial of the CEOs of the oil companies operating in Canada's Athabasca Tar Sands was live-streamed online. Polly's book *Eradicating Ecocide: Laws and Governance to Stop the Destruction of the Planet* won The People's Book Prize 2011/12. Polly Higgins sadly passed away in April 2019, but her incredible legacy lives on in the Stop Ecocide campaign to create a legal duty of care for the Earth.

In 2005, I was standing in the Court of Appeal in the Royal Courts of Justice in London, representing a client. He had been ill-treated and harmed.

This was the final day of a case that had run for three years and there was a moment of silence before the judges returned with their decision. I looked out of the window and all was still. I heard a thought in my head: 'The Earth is suffering ill treatment and harm also, and no one is doing anything about it.' This hurt me to the very core of my being. I looked out again and this time I heard the words, 'The Earth is in need of a good lawyer.'

That thought would not leave me. It was a calling. Could I represent the Earth in court? Of course I could – providing new laws were put in place. Why was it that, despite all the laws, we were destroying the Earth? If this were the case, existing law was clearly not fit for purpose. That got me thinking: could it be possible to create a universal law that protected the Earth? The way I saw it, if we destroy our Earth, we destroy our lives. The two are inseparable.

My friends said I was mad – laws for the Earth were unthinkable. What's wrong with existing environmental laws? Well, I said, they're not working – they can't be, otherwise we would not be losing the Amazon forest, we would not be destroying a vast tract of land in Canada the size of England and Wales just to give us some fuel for our cars instead of inventing new solutions. The first time I saw a picture of the Athabasca Tar Sands in Canada it stopped me in my tracks – it was an image I shall never forget. I saw a land as far as the eye could travel that was desecrated, and no one was speaking up about it. This to me is a miscarriage of justice.

We are all heroes of our fate. We all have the choice – to stand up and make a difference or to sit back and do nothing. I chose at that moment in court to stand up and give voice for the Earth. Instead of representing human clients, I decided to make the Earth my client. And if the laws did not yet exist to allow me to represent the Earth in court, then I would work out what was needed.

Amazing things happen when we commit to a course of action. That one thought led to another and set in motion a whole chain of events leading to an outcome that, back then, I could not have even imagined.

A big idea started to take shape: what if a Universal Declaration could be created? After all, we humans have a Universal Declaration of Human Rights – surely the Earth has her own rights too? My idea was that we could create a Universal Declaration that set out rights for the planet. The right not to be polluted, the right to life – these and other rights and freedoms were my starting point, so I decided to take time out to work out how to do it. I needed to go up a hill and be with Nature, to gather my thoughts and think this through, away from TV, radio and phone.

I spent two weeks with my husband up in Argyllshire in Scotland, in a remote spot I know well, where there are standing stones and ancient cup and ring carvings that look like our modern Wi-Fi symbol. We walked in the hills, brainstorming about what could be done. We envisioned a legal document that honoured the Earth's right to life – what would that entail? What rights and freedoms would we grant the Earth? My big question was: how on earth do I get the United Nations to adopt a Universal Declaration of Planetary Rights?

When I asked that question I was up a remote hillside. I had no idea how to get my idea to the United Nations. A few days later I found my mobile in my pocket and wondered if there was any chance of getting a signal, in a place where no signal seemed possible. So, I switched it on and suddenly it rang – and to my astonishment it was the United Nations. They were calling to invite me to speak about Women and the Environment, but I said that I wanted to speak about something else: rights for the Earth and my idea for a Universal Declaration. I begged the person who had phoned to ask whoever they could at the top if I could do that instead. Fifteen minutes later, she phoned back to say that HQ were very excited and wanted me to speak elsewhere.

The upshot was an invite to speak at a conference about my proposal. I dropped everything. I had five months to prepare, so I treated it like a legal brief. I spent all my time researching and preparing. By 6 November 2008, the day I was due to speak, I knew my case inside out.

The day was one I shall never forget. I felt excited at the prospect of putting my idea out to the wider world; at the same time it was a moment when it could all fall flat. Speaking to an audience of non-lawyers was quite different from speaking to a judge in court – far from the world I knew. What if they did not understand what I was talking about? It was important to me that I had to speak from the heart as well as from the head – not something lawyers are trained to do. Lawyers usually only talk about facts and evidence, not about the sacredness of life. But for me, I was going back to the original meaning of what it meant to be a lawyer, to be a 'healer of the community's woes' – only this particular community was the whole of the Earth community and all who live here.

It was only ten months before Bolivia decided to run with the proposal, by which time a group of lawyers had been involved in drafting up the initial document. Bolivia then took it to the next stage and it evolved into the Universal Declaration of the Rights of Mother Earth. Others will now continue with the process of its evolution.

But rights only get you so far. Our human right to life is protected by the crime of homicide. Where more than one is being killed, the crime is genocide. The state can prosecute someone when they kill a human, but what is the crime when someone dishonours the Earth's right to life? In a moment of clarity I realized we also need an international law of ecocide.

That thought was a light-bulb moment. I ran home and buried deep. Could this work and, if so, how? I researched and discovered that the word itself has been around for forty-odd years, yet it had no legal definition. Incredibly, I discovered that it was already a crime in wartime to destroy the environment over a certain size, duration and impact. During war there are global laws to stop mass destruction, but during peacetime you can do it all the time. Ecocide was the missing 5th Crime Against Peace – we have Crimes Against Humanity and Genocide to protect our human rights, as well as War Crimes and Crimes of Aggression to prohibit crimes that destroy humanity. Now we can put one in place to prohibit the destruction of Earth.

But there's more; restorative justice is also a large part of what needs to be brought into being – that and other Earth laws, which have yet to be written.

So much has still to be done to create the laws that put people and planet first. My ultimate vision is a world without law, one where we all take individual and collective responsibility. To get there needs a bridge – and Earth law is one such bridge. When we put in place laws that uphold the sacredness of all life, our global consciousness will shift. Law is a very powerful tool – a tool that can be used for good and bad.

Earth law are laws that place life at the very heart of all that we do.

Did you know that the word 'economics' means good household management? '*Ecos*' or '*oikos*' is ancient Greek for the home, '*nomos*' is management. Economics is all about how we manage our homeland. That really amazed me when I first heard this – a good economic system is one that cares for its homeland. However, our current economics are governed by laws that put profit first; it is the law to maximize profit, regardless of how much damage and destruction is caused along the way.

That will change when we put in place an international law that places a legal duty of care for people and planet first. That is what a law of ecocide will do.

But law in itself is only part of the solution; I firmly believe that leadership of a new kind is also required. Leadership that is bold, moral and courageous. It will take leaders who are prepared to stand up and say: 'We must stop.' Even when they are benefiting themselves from the ecocide...

One man who spoke up two hundred years ago helped change the course of history. He is not well known, but in his day he was one of the most powerful men in the world. His name was Charles Grant. He was the CEO of the British East India Company, which owned over half the world's slave trade. He stood up and said: 'This must stop.'

This man is my hero. Slavery was enormously profitable, and Charles Grant stood to lose a fortune by speaking out against it. Despite that, he did – and everything changed. People stopped in their tracks and listened. He gave voice to a thought that many others shared – and when that happened, others from within the slave trade also began to speak out. The moral imperative, he said, trumped the economic imperative. One of the most successful slave traders in the world spoke out – and when he did, governments listened.

What happened next was in itself remarkable: a law abolishing slavery was put in place very fast. People throughout the world suddenly woke up to the idea that it was no longer acceptable to treat a man badly because he was black. Instead of it being the norm, very quickly slavery became the exception.

Charles Grant was a man who showed remarkable ability to stand up and weather that storm. He demonstrated bold, courageous and moral leadership at a time when to speak out from within the ranks was unthinkable. He did it and he helped change the course of history. There are Charles Grants out there today; all it needs is for one person to stand up and say or do the unthinkable and a sea of change can occur very fast.

Each of us has the potential to be a leader. When we stand up and say 'Let's make it happen', we are demonstrating leadership. Often the person who speaks out and has a vision of what the future can hold has the skills to help bring the change into being. All it takes is commitment and help from those around us. Sometimes we meet with hurdles and stumbling blocks, and that is part of the journey. There was a time when I sat back and observed life, until it got to a point when to remain silent was no longer an option. For me, to do nothing is to remain complicit in a system that I know can be changed. And here's the wonder of it – change when it comes always happens very quickly. The Berlin Wall fell in a day, genocide was made a crime just three years after Second World War, apartheid was abolished. Those are moments that change the course of civilization overnight.

Locking up people is one response to ecocide; another is to open up a space for dialogue to begin between parties who have been harmed and the decision-makers whose decision led to the ecocide. Restorative justice is a necessary component of how we make good the damage done. This can be done in many ways, whether it be through court orders or through agreement of the parties. One thing had bothered me: what I did not want to do was create a law that had every CEO imprisoned. Yet imprisonment is a powerful disincentive, unlike fines. How would this play out? A conversation about this led to a very good idea – to hold a mock trial, with a real judge, jury and legal teams. And so it came to be: in September 2011 a high-profile one-day event was staged in the UK's Supreme Court. The charges of ecocide were real, based on the BP Gulf oil spill and the destruction of the Athabasca Tar Sands. The only people who were not real were the CEOs. What we did was to hold a trial based on real ecocides and test the law. *Sky News* live-streamed it across the world and it received worldwide news coverage.

Our CEOs were found guilty on two of the three counts. The mock trial had given us a chance to put to the test a law of ecocide and we saw that it would work. But how to sentence our CEOs? Could we bring restorative justice into the boardroom? We adjourned sentence over to a restorative justice hearing where the CEOs were given the opportunity to meet with people who represented the different human and non-human beings who had been harmed – the birds, the earth, wider humanity, future generations and the indigenous community. The aim of the sentencing was to see whether justice could be met by offering a restorative justice process. One CEO declined to participate, the other accepted. Restorative justice, when consensual, can be a transformative experience – not only for the convicted, but also for the injured parties. In this case it was a success – some solutions were agreed. As for the other CEO – the judge sentenced him to prison.

A restorative approach to justice attends to everyone's loss that arises out of the crime – and in the case of ecocide, that can be complex. What the mock trial and sentencing demonstrated is that complexity is no bar to justice.

Earth law is evolving in many ways and many more people are now engaging in how we rebalance the scales of justice, so that we re-establish intrinsic values at the very core of our lives. Instead of supporting systems that are destructive, we can put in place systems that are life-affirming. Instead of supporting laws that are creating silent rights (the right to destroy, the right to pollute), we can put in place laws that create the right to life.

My dream is for peace. I want for the violence to end. Ecocide is a violence against human and non-human beings. Destroying communities is a cultural as well as an ecological ecocide – a community that can no longer hunt or fish on its land due to pollution and destruction has been subjected to cultural ecocide. That is a violence based on an imposed value – a price-tag mentality – that fails to look to the intrinsic value, the sacredness of life itself. It is also a violence that fails to look to the consequences – consequences that can have huge ripples of effect right across the world. It is a violence that fails to honour the interconnectedness of all life.

Just as we humans have the right to life, so too does the Earth. There are other rights, too; the right not to be polluted, the right to be restored – these are rights that belong to us all, to the seas, the trees, even the humble bumble bee. Imagine that the Earth is a pot of gold: what we have been doing is taking out all the gold. Very soon we will fight for the last remaining gold coins – unless we stop. Close the door once and for all and end the era of ecocide. Do that and we can begin to put back in. That way, we can live off the interest without taking the capital. Instead of squandering the last of our gold, we can invest in it so that it keeps on growing. When we do that, we build economies that are resilient and strong. When we do that, people will become happier and we will no longer fight over dwindling resources.

When we do that, we will have peace. That is a legacy worth giving my life to.

Earth law is an idea whose time has come. We can halt the march of destruction in its tracks. All it needs is for one leader to say: 'Let's make it happen.' In 1999 Lloyd Axworthy, the Canadian Minister of Foreign Affairs, was President of the UN Security Council. He said, 'Let's ban land mines in a year.' Everyone said he was mad – but you know what? He did it, and he is a hero. This is what is needed again.

Just because something is the norm does not always mean that it is right. Two hundred years ago, slaves were a part of everyday life. People thought it was acceptable to keep Black people in chains and sell them as commodities. That was the norm until we outlawed it. Same with apartheid. Same now with ecocide. Each time, the moral imperative trumped the economic imperative, and each time we adapted. What is needed now is adaptive leadership – one that can adapt fast, a world that puts care of people and planet at the heart of its decisions, a shift from command and control to empowerment of each of us.

Sometimes a moment in time can have enormous impact on our lives – and for me, that day in court when I looked out of the window changed my life. It is driven by my vision for a better world – a world where peace does exist, a world that honours life itself, a world based on the sacredness of all life, a world where we all live in greater freedom, a world where love is all.

Further Reading & Useful Websites

Eradicating Ecocide by Polly Higgins

(2010)

Earth is Our Business: Changing the Rules of the Game by Polly Higgins (2011)

www.eradicatingecocide.com
www.pollyhiggins.com

CAROLINE LUCAS

Green Politician, UK

Born in Worcestershire in 1960, Caroline Lucas has twice led the Green Party of England and Wales and has been a Member of Parliament since 2010. She is a passionate campaigner on the environment, social justice and human rights. Caroline's career as an activist began with CND protests as a university student, and she has since worked with numerous NGOs and think-tanks, and has also campaigned and written extensively about the climate emergency, green economics, trade justice and food. A former MEP, Caroline was the first Green Party MP elected to the House of Commons in 2010. The core belief of Caroline Lucas and the UK's Green Party – the party of hope and radical change – is that fair is worth fighting for, and Caroline has worked both inside and outside parliament to champion a transforming green alternative to the long years of 'politics as usual' that have resulted in entrenched inequalities and environmental breakdown.

Opposite my desk in my Parliamentary Office hangs a copy of the front page from the Brighton *Argus* of 7 May 2010. I have my arm aloft in triumph; above is the banner headline 'History is made'. I can remember the moment vividly – the culmination of more than twenty years as an activist in the Green Party, and of the party's own journey from a few visionaries meeting in a pub in Coventry in 1973, to securing our first seat in the British Parliament.

Yet this moment also feels unreal, because as soon as you have scaled one peak, another is there before you. In this case, it was to find a way of using this breakthrough into national politics to advance all the causes and issues that matter most to Greens, while also representing the people of Brighton who had put their trust in me. So that victory night feels now like the odd, uneasy calm they say you find at the eye of a hurricane.

This experience is not unique. Campaigning brings few high points. Victories are rarely instant, or total. It's much more common to feel that perhaps you are making a little progress, that a few minds have been opened; or that an evil has been tamed for a while. And even when you do succeed, it's common for campaigners to find another cause to support. Trying to change the world is more of a compulsion than anything else – a feeling that something is wrong and that you cannot live with yourself if you do nothing about it. For me at least, the greatest satisfaction comes from the shared endeavour, the sense of common purpose with others to try to reach your goals.

So it was in Brighton. One of the characteristics of the British electoral system is that it is virtually impossible for a new party to win seats in Parliament. In countries that have forms of proportional representation, there is often a threshold or cut-off point to exclude 'fringe' parties – say 5 per cent of the vote. But in Britain, where 'winner takes all', you can win 20 per cent, even 30 per cent of the vote and still not win a seat. The only possible route to victory is by working away locally in individual constituencies, listening to people's everyday concerns, helping to give them a voice.

It may seem a long way from 'saving the planet', but unless you win people's trust street by street, community by community, you cannot expect a majority to trust you with representing them, first in local government and then in Parliament.

At the same time, you need to reach those voters who take little or no interest in local affairs. There, you need to get access to the media – to take part in debates, speak out on issues, and build up credibility. It takes commitment and discipline over many years to achieve this. It's not easy to blend this with the passion and fire, the sense of mission and excitement that campaigns need to have to attract supporters and to keep the movement going forward. Perhaps this is why no new national party had broken through into Parliament for more than sixty years.

My own involvement in this mission came about by an indirect route. I come from a very conventional background. My father ran a heating business and an ironmonger's shop, and my mother stayed at home to bring up three children. The only newspaper in the house was the *Daily Mail*, and we would never have dreamed of discussing politics around the dinner table. I owe so much to my parents, but radicalism came from elsewhere.

I was fortunate that, growing up in the 1980s, campaigning was not seen as a fringe activity but something that people from all parts of society were involved in. At the time, the British government was spending huge amounts on new nuclear weapons, deliberately raising the temperature of the Cold War stand-off with the Soviet Union, and so I joined the Campaign for Nuclear Disarmament. We protested at the two air bases where these weapons were to be deployed, Greenham Common and Molesworth, and I stood alongside students and grandparents, CND stalwarts and local mothers at their first demonstration. The issues at stake were far greater than any divisions of class, or background, or accent.

There were other causes that I supported – fighting discrimination, international development, environmental degradation – and at times it was hard to know which pressure group to back. Their campaigns provided a fantastic way to concentrate on single issues and to work with like-minded people; but they could also leave you wondering if you were only tackling the symptoms, not the causes.

In 1986, I read a book by Jonathon Porritt called *Seeing Green*, which changed all that, and changed my life too. It set out the ways in which all these issues are connected by the underlying political process – not just in national parliaments, but in international agreements and through the actions of multinational corporations in banking, the media, natural resources or manufacturing. For the first time, I could see that the decisions made – often in secret – determined the world we live in, and also how through politics, we could challenge those decisions.

The Green Party offered me a way to seek to influence the world through politics in a way that recognized these underlying connections: how our economy was built on the use of finite resources in other countries; how this led to the exploitation of the people of those countries, interference in their political freedom, even military commitments and involvement in wars; and how those who made their money and gained their power through such economic relationships were well placed to influence decision-making behind the scenes.

Part of the excitement in exploring these patterns was the way in which the decisions of political leaders became more explicable. When our government acted in ways that seemed morally bankrupt, such as propping up dictatorships, or utterly perverse, such as putting farmers in the developing world out of business by dumping food surpluses on their markets and then providing international aid when those countries could no longer feed themselves, the reasons for their actions became clearer – if still unacceptable.

From the start, Green politics was for me much more than 'environmental' issues, such as energy efficiency or Nature conservation, important though these are.

It was about recognizing that our way of life – in the rich 'North', and increasingly in developing countries as well – was unsustainable. In other words, we could only carry on with our lifestyles by taking from others – either in our own country, or around the world, or from future generations – and that even if society were based on inequality and exploitation, this would only postpone a day of reckoning. In this sense, ecological politics is more hard-edged than conservatism, liberalism or socialism. Greens see that the laws of Nature and of physics cannot be changed by wishful thinking. If you take too many fish one year, they will not be there the next. If you concentrate food production on single crops, you risk blight and famine. If you depend on scarce resources looted from other countries, you may end up having to fight for them. Greens see that actions have consequences.

Traditional politicians talk about the environment. Sometimes this is a ploy; sometimes they genuinely care, and even take action on specific issues. But these advances, however welcome in themselves, rarely address the roots of the problem. One of my first campaigns after being elected was to make the government implement its pledge to crack down on illegal timber imports. Yet behind that specific issue there is a 'back story', which ranges from corruption of politicians in developing countries by multinational traders, the violent suppression of local protest movements, the disenfranchisement of local communities and indigenous people, even the undemocratic nature of the world trade system. Tackling all these factors would be beyond the remit, or capacity, of a single NGO: but lasting remedies will only come about through a direct challenge to the political system.

That led me to the Green Party – and to sticking with it during years when it seemed the aim of a new politics was an impossible dream. I worked with talented colleagues – but a supreme effort to reach out to voters and connect to their own underlying feeling that 'something was wrong with politics' could bring in over a million votes in the European elections of 2009, yet leave us with no representation at all. In keeping up my own morale, I was helped immensely by the faith of others, first in the analysis that our work would one day help every other campaign that wanted to challenge the status quo; and that, citizen by citizen, we could win the arguments and overcome the barriers placed in the way of any new political party.

And those barriers did provide us with a crucial lesson: because we could only gain a seat in Parliament by winning at least a third of the vote, we always had to think not only about the hard core of the campaign, but the wider public who didn't know us, or what we stood for. This wasn't just about forcing ourselves to preach to the unconverted, uncomfortable though that can sometimes be. It was also about adapting to a broad audience without compromising our principles. Every campaign I have ever been involved with has to some extent faced this challenge: how to balance commitment to the cause with an understanding of how your cause looks to people from the outside, and how those outsiders can be persuaded to listen, engage, and even join. Put that way, it sounds easy – but telling the difference between sticking to your principles and an inward-looking purism is difficult.

It's a question that I've faced directly, having served the Green Party both as a principal co-speaker and as the first elected leader. When I joined the Greens, it was a core article of faith for many that we should not have a leader in the traditional mould. Formal leadership brings with it a lot of risks. Presenting one image as the 'face' of a party can attract some people but also make others feel it's not for them. At a deeper level, if leadership is about concentrating power in the hands of a single person, it is damaging to the cause and can lead to poor decision-making. But leadership is a powerful tool that can draw people in, inspire them, and help focus their energy and creativity. People also want to see the human face of an organization, to help them understand what it is there for and judge if they trust it.

For me, a telling argument was the number of potential supporters who felt that, in sticking to our 'no leader' position, we were sacrificing our effectiveness. I also believed that we could show that leadership did not have to be about seizing power and holding on to it at all costs. Leaders must have confidence in their abilities as well as their cause; but they should also be eager to share out the responsibilities of leadership with others, from shaping policy to maintaining morale, so every individual in the movement feels a personal responsibility.

We made the change, and the party is stronger as a result. I think this is because the debate within the party wasn't about sacrificing a principle to gain some votes, but about changing the way we worked to help put our principles over to more people. It is an example of how we found the right balance between principles and pragmatism; and how as a party we had become more self-confident.

Since being elected to Westminster, I have been plunged back into a whole series of campaigns and issues. It's been a chance to put into practice my belief that through politics you can support a range of causes. Just before my maiden speech, for example, I was passed some confidential documents by Greenpeace about the dumping of toxic waste in a developing country. At the time, the media were reluctant to cover the story for fear of legal challenges, and it felt like just the kind of issue that might not have been raised in Parliament had I not been elected.

Since then there have been hundreds of opportunities – in debates, in tabling Parliamentary Questions or tabling amendments to legislation – to advance causes dear to my heart, as well as the immense workload that goes with representing a constituency like Brighton Pavilion, which has its full share of social problems. I still have this sense of there being far more causes than hours in the day to support them, and I try to concentrate on those where other MPs may be less committed or are prevented by their party whip from becoming involved.

Of course, fighting the cuts, tackling unemployment or challenging the greed of city bankers matters just as much to my constituents as it does to Green Party members or supporters. As I attend meetings in Brighton or just chat to residents on the streets, I'm never told to concentrate only on local issues. More and more, the public see how their own experiences – anything from being unable to raise funds from the banks to expand their business, to being pushed into debt by blunders in the Benefits Agency – are linked to the skewed priorities of the ruling elite. Slowly, Britain is becoming radical once more.

When I was elected, I decided that in all the competing responsibilities of being the first Green MP, I would try to do one thing that truly mattered for the people of Brighton. I was struck by how the city suffered from the blight of hard drugs – being dubbed the 'drugs death capital of the UK'. Drugs is a field where the evidence often points one way and the rhetoric of politicians and the media points another. Trying to lead a crusade would be bound to fail; but working behind the scenes with police, local authorities, the National Health Service, with drug users and with their families and friends is helping to build trust and find approaches that could make a lasting difference. There are alternatives to our current national drugs policy. If I thought that leading marches or sit-in protests would change it, I would – but here at least I believe that there is a chance to persuade the establishment to try to see addiction not as a crime but an illness, and have regimes for policing and treatment that reflect this.

This is an example of the ever-present tension between radicalism and compromise.

When I am in the Chamber at Westminster, surrounded by all that grandeur and history – and by the dead weight of tradition and cosy self-satisfaction – it's hard not to want to shout out against it. In so many ways, Parliament fails to represent our country, and perpetuates the worst in our history, from class rule to dreams of Empire. It is still a club for the Establishment, and still too many MPs put personal or party advantage ahead of the interests of those they are supposed to represent. Until this is challenged, every other change we might want to see – from action on climate change to justice for the Chagos Islanders, from reducing our stocks of nuclear weapons to giving those caring for ill relatives proper support – will be fought all the way.

If we are to change the way we govern ourselves, so that it genuinely reflects people's real concerns, then we need to win over many millions more supporters. Most people in Britain know that something is very wrong with our politics – but most of them do not have a clear sense of the changes we need or how to bring them about. Nor do they feel as passionately about these changes as those who are committed to the cause.

I understand that. Having grown up in an ordinary household, with parents who did not rush off to political meetings or chain themselves to tanks, I know that not everyone is kept awake by thoughts of famine or dwindling natural resources. But I also know that those who are not engaged still care. They show this in different ways – through civic duty, or giving to charity, or helping out their neighbours. If we can reach them, help them to see the alternatives to our current way of life, they will join us.

As campaigners, our values and desires are not, at heart, that different from anyone else's. We want people to be happy, to have a chance to lead fulfilling lives, and to know that, if things go wrong, others will come to their aid. In return, people should take responsibility for their own actions, and treat others with decency and respect. Nothing very radical there.

I've seen this in Brighton. We were told you couldn't win a seat on a radical platform, and that to be elected we'd have to compromise our principles. Not so. We've won support in deprived and affluent areas alike, among students, pensioners, business people. The idea of a better world is not just for dreamers, or activists. It's for us all.

Further Reading & Useful Websites

Seeing Green: Politics of Ecology Explained by Jonathon Porritt (1984)

Alternatives to Globalisation: A Manifesto by Michael Woodin and Caroline Lucas (2004)

www.carolinelucas.com
www.greenparty.org.uk

BILL MCKIBBEN

Environmental Writer & Educator, US

Born in Massachusetts in 1960, Bill McKibben has been described by the *Boston Globe* as 'probably the US's leading environmentalist' and by *Time* magazine as 'the world's best Green journalist'. He first wrote about climate change in his seminal book *The End of Nature*, published in 1989. In 2007, Bill McKibben set aside his quiet life as a writer to found 350.org. This global grass-roots organization works to halt and reverse climate change. In 2009, Bill led 350.org's organization of the first International Day of Climate Action with 5,200 simultaneous demonstrations in 181 countries.

Bill McKibben was awarded the Gandhi Peace Award in 2013 and the following year biologists named a new species of woodland gnat – *Megophthalmidia mckibbeni* – in his honour.

I am a writer by trade, and for a long time I thought that writing would be my singular contribution to the push for a workable planet. In 1989, I published the first book for a general audience about global warming. I was twenty-eight; my theory of political change was that people would read my book, and then they would... change. And people did read the book, in numbers as large as any serious book can hope to be read: *The End of Nature* eventually came out in more than twenty languages and was a bestseller in many of them.

As it turns out, however, that's not exactly how political change comes about. For much of the next decade I kept writing more books. They didn't address climate change directly, not all of them, but they were about the same general theme: why exactly weren't we tackling the biggest problem humans have ever faced? The best of them, perhaps, was *The Age of Missing Information*. I watched every minute of a day's worth of television on what was then the planet's largest cable system (about twenty-four hours, to be exact). I was looking for the Rosetta Stone to explain why we weren't acting. And it provided plenty of clues – the hyper-individualism of a high-consumer society clearly lies near the heart of this mystery. Understanding the problem, though, was bringing me no closer to the solution – which seemed more important to aim for, the longer I wrote about the issue and the more I understood the toll it was already exacting. I remember visiting Bangladesh, experiencing its first big outbreak of dengue fever, a mosquito-borne disease spreading like wildfire across the global south. (If you were looking at our planet from a distance and speculating about why we were dramatically changing our atmosphere, a reasonable guess would be that we'd embarked on a serious mosquito-ranching experiment.) I got bitten by the wrong mosquito and was pretty damned sick myself – but the real problem was being forced to watch other people die and realizing that, as representative Third Worlders, they'd done exactly nothing to cause the problem. When the UN tries to measure how much carbon each country emits, you can't even get a number for Bangladesh; the US, by contrast, churned out about a third of the twentieth century's CO^2, carbon, which will linger in the atmosphere for an average of a century.

Somehow it seemed less and less okay to just write and speak about global warming, and more and more necessary to actually try to do something about it. But how? I mean, I'm a writer; I live in the woods. I had no clue. So I called a few of my writer friends around Vermont on the phone, and I said, 'Here is the plan: we are going to go to the federal building in Burlington. We will sit in on the front steps. We will get arrested and there will be a little story in the paper. At least we will have done something.' And these writers were as clueless as I was: 'Oh, that's a very good plan. Let's do that.' Until happily, one of them called up the police to ask, 'What will happen if we carry out this intrepid stunt?'

'Nothing will happen, stay there as long as you want.'

So we decided to recalibrate. I sent out emails to people and said we are going to go for a walk. We left a couple of weeks later from Robert Frost's old summer writing cabin up in the Green Mountains. And we walked for five days and we slept in farmers' fields at night. I am a Methodist Sunday-school teacher so I called the churches en route to make sure there would be pot luck suppers available as we walked. We got to Burlington after five days and there were a thousand people walking. In most places one thousand people is nothing. But in Vermont, except for University of Vermont ice-hockey games, that is about as many people as you get in one place at a time.

It was good. It got everybody running for office that year, 2006, to come down and meet with us at Lake Champlain. Not just meet with us – they all signed this piece of cardboard we had been carrying, saying that if they were elected they would work to cut carbon emissions by 80 per cent by 2050.

We were pleased. The only sad part was to open the newspaper the next morning and read this story that a thousand people may have been the largest demonstration on climate change that had taken place in the US. I read that and thought – it almost sort of finally clicked – 'No wonder we are losing!' We have the superstructure of the movement: we have Al Gore, the scientists, the engineers, the policy people, a million work plans. The only part of the movement we forgot was the movement part. There is nothing there to give it any heft.

We decided to see if we could do something about this. When I say 'we', I mean me and seven undergraduates at Middlebury College in Vermont. We didn't have any money, but we just started sending out emails saying: 'Do something like this [i.e. a similar demonstration] in the spring.' We picked a day in April. Sort of to our surprise, people really responded. That day in April we had fourteen hundred demonstrations across the country, and two days later both Hillary Clinton and Barack Obama, who were running for president, adopted the goal we set of an 80 per cent reduction in carbon emissions by 2050, which at the time was very radical. But they did. We were pleased – the technical word would be smug. We were quite happy with ourselves.

The problem was the summer of 2007. Six weeks later is really when all hell began breaking loose on the planet. The Arctic really started to melt for the first time in the summer of 2007. I spent the whole summer getting phone calls from scientists I had known for a quarter of a century. These guys had always been worried and sober about our prospects, but now they were panicked: 'Things are happening way faster than we thought they were going to. How can we get the word out?' By the time the summer was over, two things were clear. One, what happens in 2050 is not really of that much interest. It is what happens in 2020 that counts, and we have to move fast. Our old goals were obsolete. The second thing was that we were not going to solve this one light-bulb at a time – and we were not going to solve this one country at a time either. We were going to have to work globally.

And that was a pretty daunting realization because we had no idea how to work globally. That is why we were pleased in a kind of weird way in January of 2008 when our greatest climatologist, Jim Hansen at NASA, and his team put out a paper that said, in effect, 'We now know enough about the world to tell us how much carbon is too much.' What they said was: 'Any value for carbon in the atmosphere greater than 350 parts per million is not compatible with the planet on which civilization developed into which life on Earth is adapted.' Stern language for scientists to use. Sterner still when you know that everywhere around the world, in Vermont and Vietnam and Versailles, right now the figure is 391 parts per million CO^2 in the atmosphere – i.e., it was effectively the final declaration that this was not a future problem to worry about down the road, it was a very present emergency.

And in many ways it was a horrifying paper that once and for all said, 'We are never going to have the Earth that we once did. We have made unalterable changes, and on an enormous scale.'

But for us as organizers trying to think about how we would organize the globe, it had a use, and that use was that now we had a number. The good thing about Arabic numerals is that they cross linguistic boundaries.

You can begin to see how you might be able to do some kind of global-scale organizing, and it was good we had that one advantage because we did not have much else. Well, these seven kids had now graduated from Middlebury, which was good because there were no more papers and stuff getting in the way. And seven was probably the right number because there are seven continents, so each took one. The guy who got the Antarctic also got the Internet – it is sort of its own landmass.

So off we set to try to organize, not that we really knew what we were doing. We did not have any money to speak of or anything, but we knew what we wanted to do and we started finding people all over the world who understood what we were talking about. Most of them were not environmentalists – they were people working on public health in their communities, on food, on women's issues, on peace, and on all the things that no one is working on in Pakistan right now because they are too busy figuring out how they are going to get tarps over people's heads.

We said that to do this – to make this work at the beginning, to try to get this important number out there – we needed to have a kind of coming-out party and push it into the planet's information bloodstream. So we picked a day in the autumn of 2009. We had no idea how this would go because, as I said, we had no real idea what we were doing at all – which may have been, in retrospect, our greatest advantage in certain ways. Somebody let us borrow a dingy office in lower Manhattan for the week before, and we did all the usual things, such as putting out press releases – but basically we were just waiting for the returns to come in, as it were, because we told everybody that we wanted to upload pictures as soon as they did something on this day.

We got a little sense that it might work two days early. We had done these training camps for young people – one in Turkey for Central Asia, one in the Caribbean, and we had also done one in Africa in Johannesburg. We brought a couple of young people from every country in Africa, financed by donations. Most had never left their country. Most had never been on an aeroplane. But they were great organizers, they understood what we were talking about, and they fanned back out across Africa.

And then we did not hear much from them because in parts of Africa the Internet is still mostly notional – you cannot just Skype people constantly or whatever – but we knew they were working.

Two days before the big day, we get a call from our organizer in Addis Ababa.

She was seventeen, and almost in tears. 'The government has taken away our permit for Saturday. They are not going to let us do this thing. So we are doing it today so they cannot really stop us. We are really sorry, we know that we are jumping the gun. We do not want to spoil it for everybody. Oh, and we have 15,000 people right now out in the street in Addis chanting "three fifty".'

It was the beginning of a quite awesome kind of forty-eight hours. I mean, these pictures just flowed in from everywhere. The next one actually came completely unexpectedly from US troops in Afghanistan, who wrote '350' with sandbags and sent a note saying, 'We are parking our Humvee for the weekend and walking.'

Before the weekend was over, there had been 5,100 of these things. CNN said it was 'the most widespread day of political action in the planet's history' because it had been in 181 countries.

We've kept up this kind of giant campaign – in fact, it's grown to the point where people in every country but North Korea are involved. And we've convinced a majority of the world's governments to back strong targets. But not, sadly, the minority of rich countries doing most of the damage. There we've made no progress – more carbon pours into the atmosphere each year.

The reason we are losing is not that we are losing the argument. We have long since won the intellectual argument – climate change is the greatest danger humans have ever faced. The reason we are losing is that there is too much power on the other side, and most of that power is financial and there is no mystery about where it is coming from. The fossil fuel industry is the most profitable enterprise that human beings have ever conducted. Eight of the ten largest companies in the world are in that business. Exxon Mobil made more money last year than any company in the history of money. And in our political system, it takes only the slightest fraction of that money to corrupt that political system to the point where you do not get good or fair results out of it. In the US, for example, the Koch brothers, the third and fourth richest men in America, made their money on oil and gas, and they were one of the biggest funders of 2010's election campaign. Of the twenty-six new Republican members of the House Energy Committee, twenty-two took money from these guys, and for many of them it was their biggest contributor. No wonder, then, that they voted against a resolution simply saying global warming was real.

The oil industry will always and forever have a monetary advantage. To beat them we're going to have to find other currencies to work in, the currencies of a movement. Passion, spirit, creativity. Sometimes we'll have to spend our bodies. In 2011 we organized the largest civil disobedience action in the US in thirty years—some 1,253 of us went to jail, protesting plans for a giant pipeline from Canada's tar sands. And at least for a little while we won – the Obama administration delayed the project, in the face of great opposition from the fossil fuel industry. It may well be a temporary victory, and at best it doesn't do much to actually stop climate change. But it does show us how to proceed in the fight against this most dangerous of industries – boldly, resolutely. And with the firm conviction that we are not the radicals here. We merely want a world a bit like the one we were born on to. It's oil and coal and gas barons who are radical: they're willing to fundamentally alter the chemical composition of the atmosphere in order to make more money. Think about that – it's the definition of extreme.

I don't know if we can win this fight. But I'm glad we've engaged it. I miss my life as a writer more than I can say, but in certain ways I guess this is a literary task. We're trying very hard to at least make sure that those who are destroying the planet know that they're destroying it – we're trying to bring what's going on to our joint consciousness as a civilization. We're looking constantly for the right images and metaphors. We may lose – but at least we're in the fight now.

Further Reading & Useful Websites

The End of Nature by Bill McKibben (1989)

The Age of Missing Information by Bill McKibben (1992)

Deep Economy by Bill McKibben (2008)

Earth: Making a Life on a Tough New Planet by Bill McKibben (2011)

'Target Atmospheric CO_2: Where Should Humanity Aim?' by James Hansen et al. (2008)

www.columbia.edu/~jeh1/2008/TargetCO2_20080407.pdf
www.billmckibben.com
http://350.org

CARLO PETRINI

Slow Food Advocate, Italy

Born in Italy in 1949, Carlo Petrini is a 'slow food' activist who first came to prominence in the 1980s when he campaigned to stop McDonald's opening near the Spanish Steps in Rome. In 1989, Carlo founded Slow Food, a global, grass-roots organization that connects its supporters through the pleasures of good food and a commitment to the community and the environment; and in 2004 he founded the University of Gastronomic Sciences to bridge the gap between agriculture and gastronomy. Carlo Petrini believes that food has been stripped of its meaning, reduced to a mere commodity, and its mass production is contributing to injustice all over the world. The Slow Food movement, brings together food communities who produce, process and distribute quality food in a sustainable way.

It's a small world: a popular saying that is actually very true, for a number of reasons. I have travelled extensively for my work, and all over the world I have met people who have a similar mindset to mine, even though their experiences are far removed from my own and they live in very different places. I have shared visions of life and approaches to life with people ranging from inhabitants of the remotest, most isolated villages on impossible latitudes, to great intellectuals who give lectures on every continent.

So, I am convinced that the power of ideas is without boundaries and, above all, is never compromised by a particular local context. On the contrary, it is precisely by learning how to inhabit our own real world that universal ideas find unexpected inspiration, precious lifeblood, surprising connections, and ultimately, make us all feel part of the same humanity, capable of determining our destiny together. And so it is that, over time, I have also realized how the world is as small as my own hometown – Bra, in Piedmont, northern Italy.

I was born there in 1949, and I still live there now. It is also where the international Slow Food movement is based, of which I am president. Slow Food was founded to counter the rise of fast food and fast life, the disappearance of local food traditions and people's dwindling interest in the food they eat, where it comes from, how it tastes and how our food choices affect the rest of the world. The movement took its first steps in Bra thirty years ago, inspired by the local breeding ground of people, ideas, territories, agricultural products, customs, traditions, political passion, windows to the outside world – but also in response to problems and obstacles.

Some might perhaps find it odd that a movement like Slow Food – which now has some 100,000 members around the world, with a network of food communities (Terra Madre) that encompasses around a million people from almost every country in the world – continues to have its head office in a small town in Italy with only about 30,000 inhabitants. It will definitely seem strange to anyone who thinks that international organizations absolutely must have offices in a major capital. And some might wonder how we can possibly manage everything from the sticks. But in my opinion, the secret lies precisely in the sticks – it's a small world, which helps you identify the potential difficulties as well as the opportunities. This gets you into the habit of seeing apparently insurmountable obstacles as opportunities to change your approach, your way of thinking, your principles – to feel, as never before, like a brother of a humanity so diverse, and almost always so physically remote, that you might never otherwise truly get to know it.

My homeland is surrounded by beautiful hills, the Langhe. Here, many winemakers are based; they produce many excellent wines, some of which are considered the best in the world. However, this wasn't the case before the 1980s. The memory of Second World War and the effects of wretched poverty were still evident; the local residents had been profoundly affected, and were not exactly rolling in money. It was old-fashioned farming country, producing a number of specialities in addition to wine; and over the years, out of the ashes of famine and poverty, it has managed to create a huge appeal based on its traditional food and wine. Today, the Langhe is a tourist destination for people from around the world and the wine producers are the driving force behind an impressive restaurant industry, with dishes and ingredients that have become undisputed Italian classics. An impressive turnaround – but that should come as no surprise. After all, if you think about it, every great culinary culture has its origins in hunger, and the most delicious and brilliant traditional recipes were generally created to overcome shortages, to use up leftovers, to preserve and extend the life of food – in short, to make the most of what little there was.

To give you an example that everyone can understand, let's remember that pizza was created in the backstreets of Naples at the end of the eighteenth century to provide cheap food for workers who didn't even have a bathroom at home, let alone a kitchen. Cheap ingredients to keep down the costs at a time of atrocious poverty, food sold on the street and eaten in a hurry – Neapolitans invented something simple but incredibly versatile, which has now become universal. And, if we're completely honest about it, they also invented something that is very similar to fast food.

Going back to Langhe, though, I think about using up leftovers, talked about so readily today when considering the merits of 'happy simplicity'. It reminds me about raviole (as ravioli is called in the Piedmont dialect – masculine in Italian, feminine in Piedmontese), a type of stuffed pasta. But stuffed with what? In restaurants today, they use the best meat and vegetables; but raviole were originally created to use up the week's food scraps. On Sunday you would take a thin sheet of pasta and the leftover meat or vegetables would be chopped up with a few herbs or some cheese to make the filling, creating new flavours and textures. I saw my own grandmother do just that, at least once a week – and it was food fit for a king.

Necessity is the mother of invention, and there is pleasure to be had from food even in less than perfect conditions. In the end, it is perhaps precisely because of these less than perfect conditions that you achieve something better, more beautiful, more beneficial. The need to eat outside the home for not much money invented the pizza; the need not to waste precious food created raviole, which can still be found in homes throughout Piedmont (although less so now), and is a must in restaurants in my region.

But let's go back to the wine from my neck of the woods. I was saying that the people who grew vines generally had small plots of land, were from backgrounds of great poverty, and usually gave the grapes to big cellars or producers – who, even though they might aim for a certain quality, failed to bring out the diversity and potential richness of the wine, which ended up comparing badly with wines from other traditional winemaking regions, such as Bordeaux in France.

I would often visit these people, these small producers, some of whom taught me a lot about life. Some of the older ones had been key figures in the resistance against the Germans at the end of the Second World War, real intellectuals like Bartolo Mascarello, bringing together in their own cellars some great thinkers and incredible characters, who had so much to teach us. The main wine produced in this area, Barolo, had a long history and, even as far back as a century earlier, some wine experts had seen its potential for ageing, for developing a complexity and richness to rival those of the best French wines. But in the 1960s and 70s, Barolo was often given away to anybody who bought a certain quantity of other wines, because it didn't seem to be worth anything; and it was hardly ever produced any more to bring out its characteristics.

Ironically, it was a huge crisis – a scandal connected to the criminal practice of adulterating wine by adding methanol to it (resulting in several deaths) – that triggered the renaissance. At the time (1986), the events that went down in history as 'the methanol wine scandal' seemed likely to sound the death knell for the nation's wine industry. Nobody from other countries wanted to buy Italian wine any more, and it was of little importance that only a few wine producers were guilty of the practice and that they were from a very contained area, outside Langhe. The whole industry suffered the very serious consequences. But it was from this crisis that the Langhe rose up. The small producers I knew and had spent so much time with, started to produce wine themselves, striving to bring out the diversity that their own land could produce. It was the first of these producers who went on to form Slow Food, a local association of people passionate about promoting Barolo.

In a microcosm like the Langhe, there was a macrocosm of different agricultural traditions, different soils, different levels of exposure to the sun (thanks to the surrounding hills) and different philosophies when it came to producing the wines. The result was and is extreme diversity, which, although resulting in wines with recognizable and shared characteristics, can also create huge differences, highlighting the individual skills of the producers. Small plots and small producers: in just a few years, this explosion of diversity marked the renaissance of Barolo and other wines from the Langhe, taking them to the most prestigious of world stages, making fortunes for the children of the people who had starved during the war. Many of these young people began to experiment with new techniques, started to travel and to apply to their own grapes and vines the methods that made wines from around the world so great, while always respecting their own traditions. This would often lead to open conflict with their fathers, who were used to doing what their own fathers and grandfathers had done for decades. But it was a great success: thanks to the desire to start again, to new ways of thinking and producing, to nurturing the diversity of a large number of small producers rather than a small number of large ones, the crisis of the methanol wine scandal, which had threatened everything, turned into an opportunity for the producers to change and grow without turning their backs on the past and their roots.

In my region, I watched the progress of my producer friends from close up. And in the meantime Slow Food was taking shape, a germ of an idea was slowly taking root – thanks also to the growth of the country's food industry, which was being spurred on by the renaissance in the wine industry.

Looking at my traditional foods and at the farming world where they are produced in the areas around my town, I have learned about gastronomy, about wine, and about promoting agricultural products. And I have learned about it together with friends, spending time with them, sharing my experiences with them. From the outset, Slow Food made 'conviviality' the basis upon which to begin and grow on a local level. But this didn't mean being closed-minded – on the contrary, we were keen to build a worldwide network. We completely agreed with what Ivan Illich wrote about conviviality: 'The ability of a human community to develop harmonious interaction with the individuals and groups that make it up and the ability to welcome that which is foreign to this community.' Thus, from the very beginning, while Slow Food was taking its first steps with restaurateurs, winegrowers and the ordinary people of the Langhe, it was also putting down roots elsewhere. Thanks to travel and interaction, the association was making friends with those who shared its basic principles – the battle against the sameness, the standardization and the artificialization of food. The defence of and right to pleasure – the motto of the movement's original manifesto – demand the protection of diversity.

So from my winegrowing friends, who had so brilliantly overcome the greatest crisis that could have hit their industry, I learned that no problem could take the wind out of the sails of regions that understand their strengths, rich with the benefits of their people, their humanity. Seeing where the winemakers were in the 1960s and where they are now, I believe that this transformation is possible wherever diversity is valued and food itself has a value beyond a label with a price on it – and this means food produced by people for other people, not by industry for consumers.

When it comes to wine and gastronomy, I know what I'm talking about. When I first started talking to groups of ecologists and environmentalists about these things, about the pleasure of food as a key to change, many would roll their eyes. Today, some products from the Langhe display such quality and such international success that to those who have never seen the area, they might appear to be elite products, only available to a rich clientele. But when they are backed up by local history, by local memories, everything changes.

Faced with that huge crisis, people's way of thinking changed – they looked at the rest of the world and started to learn new things, applying them to their own land to improve their diversity without distorting it. They changed their principles.

So towards the end of the 80s, the Slow Food movement began to grow and spread through Italy and the rest of the world. The idea of implementing material culture, of pursuing the guaranteed pleasure of traditional cooking, encouraged us to forge ties in all areas, to find out about other people's traditions, to travel in order to add to our knowledge. We soon discovered that the background of all the best agricultural products (and so foods) was marked by human difficulty and poverty, and those who could not manage to free themselves from such a past ran the risk of being overwhelmed by the standardization of industrially produced food.

Over the years, various projects have been established to protect biodiversity in the food industry. The Slow Food

Foundation for Biodiversity – 'Presìdi' – supports small groups protecting plant varieties, animal species, and ways of processing food that are under threat. If you like, the 'Presìdi' represents another way of reacting to a crisis: a resistance to the deluge of standardization. This work has introduced us to farmers and producers – from throughout Italy first of all, and then throughout the rest of the world. It has helped us appreciate how their practices, facing severe threats, actually incorporate all the 'antibodies' needed to deal with the major crisis now revealing itself with all of its power all around the world – the threat to established roots, local economy, the memories of traditions and sustainable expertise; the need to create new distribution channels for food; and the need to find a way of working with the Earth and its resources that respects rather than over-exploits it, by using the best possible methods and – like making raviole – getting the most out of the least.

We wanted to protect the biodiversity that guaranteed the best products and we soon learned that this was intrinsically connected to the human context of the area in which it developed. Alongside what Nature has to offer us through food, a culture has always blossomed. And this cultural diversity was as much under threat as the biological diversity. But only one thing can guarantee the survival of them both: the involvement of real people in the campaigns all around the world. The human beings who live for the Earth, who work the earth to produce food, are always the first to be threatened. It is no coincidence that in the countries in the north of the world, the size of the workforce in the countryside has fallen to an all-time low, and in some cases has all but disappeared. If there is nobody in the countryside, 'human' food cannot be produced, diversity cannot be preserved, the greatest creative strength we have to deal with the many crises that are encroaching on us more and more intensely will be lost.

There are many chapters in the history of Slow Food, some of which seemed insignificant but turned out to be decisive, and this is not the place to go through a story as long and complex as that of the people involved in the movement. We started out by defending the right to pleasure for all, before moving on to defending biodiversity and gastronomy as the greatest interdisciplinary expression of the history and fortunes of humanity (from biodiversity to cultural diversity), and so of reality; and we ended up by creating the largest worldwide network of small communities committed to sustainable food – the Terra Madre network, a network whose main resource is the people and communities that belong to it.

Terra Madre began as a large meeting of farmers, fishermen, artisans, nomads and producers of sustainable food from around the world – a meeting that took place for the first time in Turin in October 2004, and changed Slow Food for good. Since then, these people (now encompassing over 2,000 food communities, represented by some 6,000 delegates from 173 countries around the world) have met up every other year in Turin. In between, they have a permanent network, with members who communicate, who connect, who encourage conversation and travel, which reaffirms their pride and dignity. Terra Madre gets involved in local areas but has a global dimension, without any centralized management, governed only by what I like to define as 'an austere anarchy driven by emotional intelligence'. The way that ideas and principles are shared keeps it united, with each member free to claim its own sovereignty over its food; but the ties of emotion, of fraternity that have been established are something new, driving us towards new ways of creating networks and critical masses. New principles that the 'Earth's intellectuals' in the community teach us.

The real strength lies in the fact that these communities are the real protagonists in their regions, because they strive to produce food that Slow Food has defined as 'good, clean and fair': good in terms of taste and culture, clean because it is sustainable at every step of the way from the field to the table, and fair because it respects the rights of human beings and of the Earth. These communities educate, practise, work, share, find new strength to overcome new difficulties – in fact, they do what we have always done, producing food while taking the context into account, responsibly, with respect for traditions and local memories. Local food, made by locals for locals. Different from one region to the next, but driven by the same spirit. These protagonists determinedly reaffirm a right that is increasingly undervalued or taken for granted – the right of human beings to food, which we can only strive for by putting food back at the centre of our lives, just like the producers do every day.

The world of Terra Madre and Slow Food – today a huge, unique network uniting producers and co-producers (because we no longer want to define ourselves as consumers, we want to be the allies of the people producing the food, understanding that 'eating is an agricultural act') – is a small world in every sense. Small because of the strength of its local dimensions, roots in its communities and 'conviviality', but also small because of the global connections that build on shared ideas, making us exclaim when we come across someone we know when we're a long way from home: 'What a small world!' A small world that makes me contemplate the world of the future. Or at least believe in a new world.

As a foodie in my own small world of the Langhe, I have seen how poverty and difficulties have historically inspired culinary creativity; how the worst possible crises, such as that of the methanol wine, can trigger a change in people's thinking. With the Terra Madre network, I see that the problems of the food community are tackled with great humanity, thanks to the ability to combine traditional skills with new technology, with new approaches backed up by past memories.

The world that claims to be large is currently facing an entropic crisis, a crisis that is not cyclical, that will not end without major changes. Financial crisis, food crisis, energy crisis, climate crisis: the dominant way of thinking, the liberalist economic theories, the illusion that mankind dominates Nature at will, have ultimately caused crises with disastrous effects on the environment, and even more so on people.

This is why I am aware of a disaffection with the democratic process, I am aware of discouragement, of dismay, and of great uncertainty about the future. While the situation continues to get more difficult, there are many reasonable calls to 'be indignant' and 'rebel': the human dimension asks for and demands a reaction. But perhaps all of this is based on an incurable pessimism, hailing change as a revolution, a reaction against what seems more than ever to be a siege of our existence.

I have thought about it a lot. I have looked at my small world of Bra, my own personal story, and the small world of Terra Madre and Slow Food: a beautiful story of community. I am coming round more and more to the idea that we don't need to be pessimistic. It is true that we are going through a critical period for the world, but I can also see active, widespread and powerful antibodies fighting this illness. I won't let myself be overcome by the sadness or depression that demand a rebellion: I take comfort in the feeling of fraternity that will no doubt lead to a long-awaited change in the way we think. Not for the first time, humanity is faced with an urgent need to change, and it has always been the case that when a crisis rules, the embers continue to burn in the ashes. I see the community of Terra Madre, the members of Slow Food, the people who lead the way in urban agriculture, the farmers' markets, the 'Thousand Gardens in Africa' (a project Slow Food is currently involved in), the countless people who are changing the world every day by choosing to put food back at the centre of their lives, even if it's just by changing one spending habit. I see local communities getting stronger, the knowledge of where our food comes from growing, an approach to education that is more and more about 'teaching food', the young people who are going back to the land. I see the changes, and they seem to be more powerful than the crisis: we just need to give them time to spread. Slowly, and without letting ourselves become burdened by anxiety – because you should remember that it's a small world, and that ideas have long legs, and can go a long way.

Further Reading & Useful Websites

The Case for Taste by Carlo Petrini (2004)

Slow Food Revolution: A New Culture for Dining and Living by Carlo Petrini and Gigi Padovani (2006)

Slow Food Revolution: Why Our Food Should be Good, Clean and Fair by Carlo Petrini (2007)

Terra Madre: Forging a New Global Network of Sustainable Food Communities by Carlo Petrini (2010)

www.slowfood.com
www.terramadre.org

VANDANA SHIVA

Ecofeminist & Environmentalist, India

Born in India in 1952, Vandana Shiva is a key figure in the anti-globalization movement, best known for her support of the wisdom and traditional practices of indigenous Indian peoples. In 1982, she founded the Research Foundation for Science, Technology and Ecology, which led to the creation of Navdanya, a woman-centred movement to protect the diversity and integrity of living resources in India and to promote organic farming and fair trade. In 2004, she started Bija Vidyapeeth, an international college for sustainable living. Vandana Shiva's organization, Navdanya, supports a number of initiatives and campaigns to help preserve the cultural and farming traditions of India.

The Earth University – Bija Vidyapeeth – was founded by Vandana with Satish Kumar following the 9/11 attacks in New York in 2001, and is run in partnership with the UK's Schumacher College. Students at Bija Vidyapeeth explore and practise the art and science of sustainability based on ecological principles at Navdanya's organic Biodiversity and Conservation Farm.

I started my life dreaming about being a physicist, with Einstein as my aspiration. I obtained my PhD from the University of Western Ontario on hidden variables and quantum theory. But the seeds of activism had been sown in me by the Chipko movement, and the tree that grew from those seeds started to overshadow the plant that was my first passion – figuring out how Nature moulds from the eye and mind of a quantum physicist.

I was born in the Himalayas and grew up there. My father was a conservator of forests, in days before roads, so we would watch or go on horseback through the forests of oak and rhododendron, devdar and pine. Before I left to do my PhD in Canada, I wanted to visit my favourite forests and streams. However, the forest had been replaced by apple orchards, the stream was a trickle. I felt a personal loss, as if a part of me had disappeared. While talking about the disappearance of forests in a little roadside *dhaba* (tea shop), the conversation brought up 'Chipko' – the new movement where women said they would hug the trees *(chipko)* to prevent them from being felled. From that day onwards, I became a student in two universities – the university in Canada for my studies in foundations of quantum theory and philosophy of science, and the 'university' of Chipko, for my apprenticeship in activism to protect our biodiversity, our forests, our rivers, our ecosystems.

Since Chipko was inspired by the legacy of Gandhi, I also got my training in non-violent, peaceful, but very determined resistance. In 1981, Chipko was successful in stopping the logging of the forests in the catchment of the Ganga above 1,000 metres.

The following year, the Ministry of Environment asked me to look at the impact of limestone mining in Doon Valley. In those days, I was doing interdisciplinary research on Science Policy at the Indian Institute of Management in Bangalore.

Doon Valley is where I was born, where my parents lived. So I jumped at the opportunity. We did a study based on participatory research – research carried out by communities impacted by an activity. While the women of the villages might not have known the hydrogeology of karst limestone, they did know where their springs were, and when the spring disappeared. They knew which mine triggered which landslide and washed away which village. In matters of life, living and survival, local communities – especially women – are the experts.

Our study led to the closure of the mines; the Supreme Court ruled that if commerce undermines life, then commerce must stop, because life has to carry on, and under Article 21 of the Constitution, the state must guarantee the right to life to every citizen.

My mother offered me her cowshed to start the Research Foundation for Science, Technology and Ecology (RFSTE) – and I left Bangalore to become a full-time activist, with research and knowledge as a key to activism. In the Cartesian-Baconian paradigm, knowledge is linked to the power of the powerful – those who control capital, and through capital appropriate and privatize our natural wealth and our heritage. I decided that through the work of the Research Foundation, we would link knowledge to the power of the excluded – show that they have knowledge, and their knowledge should count.

In 1984, India had two major disasters. The first disaster was Punjab, the land of the Green Revolution. Extremism was exploding and 30,000 people were killed. Finally, the army was called in and they entered the Golden Temple to capture Bhindranwale, the leader of the extremist movement. The vicious cycle of violence took the life of India's Prime Minister, Indira Gandhi.

The second disaster was Bhopal, where a Union Carbide pesticide plant leaked a deadly gas, killing 3,000 on the night of 2 December. Thirty thousand have died since then, and hundreds of thousands are crippled for life. By the end of 1984, my mind was spinning. I kept asking why agriculture had become like war, and decided to do research to find out. The outcome was my book *The Violence of the Green Revolution*, which I wrote for the United Nations University. I realized that industrial/chemical agriculture was like war because it came from war. Fertilizers came from explosives factories, pesticides were war chemicals. And I committed myself to creating a non-violent agriculture for peace.

In 1987, because of my book on the Green Revolution, I was invited to a conference on biotechnology where the same companies that had brought us war chemicals, which they transformed into agrichemicals, now wanted to genetically modify and own the seed through patents. I realized we had to defend the integrity, diversity and freedom of life from these new threats, taking inspiration from Gandhi's spinning wheel. I started to save seeds, a movement that grew to be Navdanya.

Navdanya is a women-centred movement and network of seed keepers and organic producers spread across sixteen states in India. It has been working towards its goal to promote peace and equality, harmony, justice and sustainability for twenty-five years now. We have a vision that all humans have a fundamental right to ecological, economic and political security, to the protection and defence of their resources, their livelihoods and production and consumption patterns shaped by people through their participation. Biodiversity provides the basis of livelihoods of the marginalized majority – of women, peasants, tribals, fisherfolk. Biodiversity offers the potential to overcome poverty, hunger and powerlessness. Biodiversity-based food production provides climate resilience and food and nutritional security.

Through biodiversity, we envision improving the productivity and incomes of rural communities, thus combining the conservation of Nature with the eradication of poverty, destitution and misery. Our twenty-five years of successful implementation is testimony that praxis is possible.

In twenty-five years, Navdanya has helped set up sixty-five community seed banks across the country; trained over 500,000 farmers in seed sovereignty (self-sufficiency), food sovereignty and sustainable agriculture over the past two decades, and helped set up the largest direct marketing, fair trade organic network in the country. Navdanya has also set up a learning centre, Bija Vidyapeeth School of the Seed, on its biodiversity conservation and organic farm in Doon Valley, Uttarakhand, North India. Thus from training, research, advocacy, to finding livelihood solutions and actual praxis, Navdanya has been a pioneer in engendering food security, water sovereignty, seed sovereignty, and land and forest sovereignty in the country, so that everyone in India can live with dignity, equality and justice.

The Bija Vidyapeeth, which we also call the Earth University, was started because Satish Kumar was very keen to see a Schumacher-like institute at Navdanya. Some of the best thinkers and activists of our times have come and taught at the Bija Vidyapeeth. They include Edward Goldsmith, founder of *The Ecologist*; Mohamed. Idris, founder of Third World Network; Fritjof Capra, who wrote *The Tao of Physics*; Anita and Gordon Roddick, who started The Body Shop;

Masanobu Fukuoka, who wrote *The One-Straw Revolution*; Tewolde Egziabher, the Environment Minister of Ethiopia; Frances Moore Lappé, who wrote *Diet for a Small Planet*; Arundhati Roy; Sunderlal and Vimla Bahuguna; Venerable Samdhong Rinpoche, the former Prime Minister of the Tibetan government in exile; and of course dear Satish, who is with us every year to teach Gandhi and globalization.

Globalization is what has shaped the last twenty-five years of my activism.

Joining with other activists such as Doug Tompkins, Jerry Mander, Maude Barlow, Mike Ritchie, Tony Clarke, Andrew Kimbrell, Sara Larrain and Martin Khor, we formed the International Forum on Globalization. When the World Trade Organization (WTO) met in Seattle, we mobilized citizens of the world to say 'no' to corporate rule, which is commodifying everything. We declared 'Our World Is Not For Sale'.

Corporations such as Monsanto, the biggest seed corporation, had hoped to patent seed and indigenous knowledge. We called this 'biopiracy'. We have fought and won cases of biopiracy for crops such as neem, basmati rice and wheat.

Our campaign against biopiracy began in 1994 when we filed a legal opposition against the USDA and W.R. Grace patent (Patent No. 436257 B1) on the fungicidal properties of neem in the European Patent Office (EPO) in Munich, Germany. Along with the RFSTE, the International Federation of Organic Agriculture Movements and Magda Alvoet, former Green Member of the European Parliament, also supported the challenge. The patent was revoked in 2000 and reconfirmed and revoked in its entirety in May 2005.

In 1998, Navdanya started a campaign against Indian basmati rice biopiracy and patent on life by the US company RiceTec Inc. (Patent No. 5663484), and on 14

August 2001 achieved another victory when the United States Patent and Trademark Office (USPTO) revoked large sections of the patent. These included (i) the genetic title of the RiceTec patent, which earlier referred to basmati rice lines; (ii) the sweeping and false claims of RiceTec having 'invented' traits of rice seeds and plants including plant height, grain length and aroma, which are characteristics found in our traditional basmati varieties; and (iii) claims to general methods of breeding, which also constituted piracy of traditional breeding carried out by farmers and our scientists (of the twenty original claims, only three narrow ones survived).

Navdanya's third victory on the Intellectual Property Rights front came in October 2004 when the EPO revoked Monsanto's patent on the Indian variety of wheat 'Nap Hal'. On 21 May 2003, Monsanto had been assigned a patent on wheat by the EPO (EP 0445929 B1) under the simple title 'plants'. On 27 January 2004, the RFSTE, along with Greenpeace and Bharat Krishak Samaj (BKS), filed a petition at the EPO, challenging the patent rights given to Monsanto on Nap Hal. The patent was revoked on October 2004 and it once again established the fact that the patents on biodiversity, indigenous knowledge and resources are based on biopiracy and there is an urgent need to ban patents on life and living organisms including biodiversity, genes and cell lines.

In the past twenty-five years, many changes have taken place. India's patent laws have been amended; a Biodiversity Act and a Protection of Plant Varieties and Farmers' Rights Act have been passed. The Traditional Knowledge Digital Library has been set up.

From Gandhi, I have learned that effective activism combines resistance with constructive work. We resist GMO seed monopolies and patents through the Seed Satyagraha – a non-violent, non-cooperation with laws that claim that seed is a corporate invention and can be the property of Monsanto. We have been effective in preventing a Seed Law that would undermine farmers' freedom to save, exchange and breed seeds. I am now working towards a global campaign on seed sovereignty.

The last twenty years have seen a very rapid erosion of seed diversity and seed sovereignty, while control over seed has been concentrated into a very small number of giant corporations. At the Plant Genetic Resources Conference in Leipzig in 1995, organized by the UN, it was reported that 75 per cent of all agricultural biodiversity had disappeared because of the introduction of 'modern' varieties, which are always cultivated as monocultures. Since then, the erosion has accelerated. The introduction of the WTO Trade-Related Intellectual Property Rights (TRIPS) Agreement has accelerated the spread of genetically engineered seed, which can be patented and for which royalties can be collected. Navdanya was started in response to the introduction of patents on seed in the TRIPS of the General Agreement on Tariffs and Trade (GATT, the predecessor of WTO), about which a Monsanto representative later stated: 'In drafting these agreements we were the patient, diagnostician, physician all in one.' Corporations defined a problem – and for them the problem was farmers saving seeds. They offered a solution – and the solution was to introduce patents and intellectual property rights on seed, thereby making it illegal for farmers to save seeds. As a result, acreage under GM corn, soya, canola and cotton has increased dramatically.

Besides displacing and destroying diversity, patented GMO seeds are also undermining seed sovereignty. Across the world, new seed laws are being introduced that enforce compulsory registration of seed, thus making it impossible for small farmers to grow their own diversity, and forcing them into dependency on giant seed corporations. Corporations are also patenting climate-resilient seeds evolved by farmers, thus robbing farmers of using their own seeds and knowledge for climate adaptation.

Another threat to seed and seed sovereignty is genetic contamination. India has lost its cotton seeds because of contamination from Bt Cotton (cotton seed genetically modified with pest-resistant Bt: *Bacillus thuringiensis*). Canada has lost its canola seed because of contamination from Roundup Ready canola. Mexico has lost its corn because of contamination from Bt Cotton. After contamination, Biotech Seed Corporation sue farmers with patent infringement cases, which is what happened in the case of the Canadian farmer Percy Schmeiser. That is why more than eighty groups came together and filed a case to prevent Monsanto from suing farmers whose seed had been contaminated.

As farmers' seed supply is eroded, and farmers become dependent on patented GMO seed, the result is debt. India, the home of cotton, has lost its cotton seed diversity and cotton seed sovereignty – 95 per cent of cotton seed is now Monsanto's Bt Cotton. The debt trap created by being forced to buy seed every year, with royalty payments, has pushed hundreds of thousands of farmers to suicide – and of the 250,000 farmers' suicides, the majority are in the cotton belt.

Even as the disappearance of biodiversity and seed sovereignty creates a major crisis for agriculture and food security, corporations are pushing governments to use public money to destroy the public seed supply and replace it with unreliable, non-renewable, patented seed, which must be bought every year.

We have started 'Fibres of Freedom' to make cotton once again the fibre of freedom that Gandhi spun by creating community seed banks, teaming farmers in organic farming and helping create fair and just markets for beautiful hand-crafted fabric.

The organic movement 'grown by Navdanya from seed to table' has shown that we can grow more food and nutrition by conserving biodiversity and building living soil. Our activism shows that you can protect Nature and also provide for human needs.

Corporate control over the Earth's resources and people's basic need is on the one hand destroying Nature and on the other hand denying people their right to work, and to food and water. And this control is institutionalized by false claims of hunger productivity and efficiency. It has been falsely claimed that GMOs are necessary to produce more food. Our report 'The GMO Emperor Has No Clothes' shows that not only is there a 'failure to yield', there is a failure to control weeds and pests. In fact, GMOs have created super-weeds and super-pests. GMOs are therefore a threat to food security.

Just as there is a technological illusion about chemical industrial farming and genetic engineering, there are economic illusions around constructs such as 'growth'.

Globalization, as deregulated commerce, was supposed to create a new age of prosperity. Instead, it has given us a deep economic crisis that has affected all parts of the world.

The dominant economic model based on limitless growth on a limited planet is leading to an overshoot of the human use of the Earth's resources. This is leading to an ecological catastrophe. It is also leading to an intense and violent 'resource grab' of the Earth's shrinking resource base – land, biodiversity, water – by the rich and powerful, without any adjustment from the old resource-intensive, limitless growth paradigm to the new reality. The only outcome for the poor will be ecological scarcity in the short term, with deepening poverty and deprivation. In the long run, it means the extinction of our species, as climate catastrophe and extinction of other species make the planet uninhabitable for humans. Failure to make an ecological adjustment to planetary limits and ecological justice is a threat to human survival. The Green Economy that was pushed at Rio+20 could well become the biggest resource grab in human history, with corporations appropriating the planet's green wealth, and biodiversity, to make 'green oil' for bio-fuel, energy, plastics, chemicals – everything that the petrochemical era based on fossil fuels gave us. Movements worldwide have started to say 'No to the Green Economy of the 1 per cent'.

But an ecological adjustment is possible, and is happening. This ecological adjustment involves seeing ourselves as a part of the fragile ecological web, not outside and above it, and somehow immune from the ecological consequences of our actions. Ecological adjustment implies that we see ourselves as members of the Earth's community, sharing its resources equitably with all species and within the human community. Ecological adjustment requires an end to resource grab and the privatization of our land, biodiversity and seeds, water and atmosphere.

It requires the recovery of the commons and the creation of 'Earth Democracy'.

The dominant economic model based on resource monopolies and the rule of an oligarchy is in conflict not just with the ecological limits of the planet, but also with the principles of democracy, and governance by the people, of the people, for the people. The adjustment from the oligarchy will further strangle democracy and crush civil liberties and people's freedom of choice. Bharti Mittal's statement that politics should not interfere with the economy reflects the mindset of the oligarchy that democracy can be done away with. This anti-democratic adjustment includes laws like homeland security in US, and multiple security laws in India.

Calls for a democratic adjustment from below are being witnessed worldwide in the rise of non-violent protests, from the Arab spring to the American autumn of the 'Occupy' movement and the Russian winter challenging the hijack of elections and electoral democracy.

We face multiple crises – the ecological crisis, including climate, biodiversity and water; the economic crisis of deepening poverty and emerging poverty; and a social and political crisis of democracy. These crises are interconnected, as is the solution. As I look ahead into the future, I see my activism guided by the paradigm of Earth Democracy based on living democracy, living economies and living culture.

To create living democracy, we have to widen our embrace to include all life on Earth, the Earth community; we have to move from representation to participation. To create living economies, we have to move from growth to well-being of the Earth and human communities. We have to move from consumerism to conservation. We have to move from privatizing the Earth's resources to sharing the commons.

To create living cultures, we have to move from greed to caring. At the heart of this transition is care for the Earth. That is why I am part of the emerging movement for the Rights of Mother Earth. On the rights of the Earth are based human rights and rights of future generations. The Earth is calling us to be Earth activists and Earth citizens. If we listen, we have a future.

Further Reading & Useful Websites

The Violence of the Green Revolution
by Vandana Shiva (1991)

Stolen Harvest: The Hijacking of the Global Food Supply by Vandana Shiva (1999)

Earth Democracy: Justice, Sustainability and Peace by Vandana Shiva (2005)

Manifestos on the Future of Food and Seed
edited by Vandana Shiva (2005)

Soil Not Oil: Climate Change, Peak Oil and Food Insecurity by Vandana Shiva (2009)

Making Peace with the Earth: Beyond Land Wars and Food Wars by Vandana Shiva (2012)

www.navdanya.org
www.fibres-of-freedom.com

GRETA THUNBERG

Climate Activist

Greta Thunberg was born in 2003. In August 2018, she started a school strike that became a movement called Fridays For Future, which has inspired school strikes for climate action in more than 150 countries involving millions of students. Thunberg has spoken at climate rallies across the globe, as well as at the World Economic Forum in Davos, the U.S. Congress, and the United Nations. In 2019, she was named *Time*'s Person of the Year. Thunberg is vegan, and doesn't fly, in order to live a low-carbon life.

SPEECH TO THE HOUSES OF PARLIAMENT IN THE UK ON 23 APRIL 2019

'My name is Greta Thunberg. I am sixteen years old. I come from Sweden. And I speak on behalf of future generations.

I know many of you don't want to listen to us – you say we are just children. But we're only repeating the message of the united climate science.

Many of you appear concerned that we are wasting valuable lesson time, but I assure you we will go back to school the moment you start listening to science and give us a future. Is that really too much to ask?

In the year 2030 I will be twenty-six years old. My little sister Beata will be twenty-three. Just like many of your own children or grandchildren. That is a great age, we have been told. When you have all of your life ahead of you. But I am not so sure it will be that great for us.

I was fortunate to be born in a time and place where everyone told us to dream big; I could become whatever I wanted to. I could live wherever I wanted to. People like me had everything we needed and more. Things our grandparents could not even dream of. We had everything we could ever wish for and yet now we may have nothing.

Now we probably don't even have a future anymore.

Because that future was sold so that a small number of people could make unimaginable amounts of money. It was stolen from us every time you said that the sky was the limit, and that you only live once.

You lied to us. You gave us false hope. You told us that the future was something to look forward to. And the saddest thing is that most children are not even aware of the fate that awaits us. We will not understand it until it's too late. And yet we are the lucky ones. Those who will be affected the hardest are already suffering the consequences. But their voices are not heard.

Is my microphone on? Can you hear me?

Around the year 2030, 10 years 252 days and 10 hours away from now, we will be in a position where we set off an irreversible chain reaction beyond human control, that will most likely lead to the end of our civilization as we know it. That is unless in that time, permanent and unprecedented changes in all aspects of society have taken place, including a reduction of CO^2 emissions by at least 50 per cent.

And please note that these calculations are depending on inventions that have not yet been invented at scale, inventions that are supposed to clear the atmosphere of astronomical amounts of carbon dioxide.

Furthermore, these calculations do not include unforeseen tipping points and feedback loops like the extremely powerful methane gas escaping from rapidly thawing arctic permafrost.

Nor do these scientific calculations include already locked-in warming hidden by toxic air pollution. Nor the aspect of equity – or climate justice – clearly stated throughout the Paris agreement, which is absolutely necessary to make it work on a global scale.

We must also bear in mind that these are just calculations. Estimations. That means that these 'points of no return' may occur a bit sooner or later than 2030. No one can know for sure. We can, however, be certain that they will occur approximately in these timeframes, because these calculations are not opinions or wild guesses.

These projections are backed up by scientific facts, concluded by all nations through the IPCC. Nearly every single major national scientific body around the world unreservedly supports the work and findings of the IPCC.

Did you hear what I just said? Is my English OK? Is the microphone on? Because I'm beginning to wonder.

During the last six months I have travelled around Europe for hundreds of hours in trains, electric cars and buses, repeating these life-changing words over and over again. But no one seems to be talking about it, and nothing has changed. In fact, the emissions are still rising.

When I have been travelling around to speak in different countries, I am always offered help to write about the specific climate policies in specific countries. But that is not really necessary. Because the basic problem is the same everywhere. And the basic problem is that basically nothing is being done to halt – or even slow – climate and ecological breakdown, despite all the beautiful words and promises.

The UK is, however, very special. Not only for its mind-blowing historical carbon debt, but also for its current, very creative, carbon accounting.

Since 1990 the UK has achieved a 37 per cent reduction of its territorial CO^2 emissions, according to the Global Carbon Project. And that does sound very impressive. But these numbers do not include emissions from aviation, shipping and those associated with imports and exports. If these numbers are included the reduction is around 10 per cent since 1990 – or an an average of 0.4 per cent a year, according to Tyndall Manchester. And the main reason for this reduction is not a consequence of climate policies, but rather a 2001 EU directive on air quality that essentially forced the UK to close down its very old and extremely dirty coal power plants and replace them with less dirty gas power stations. And switching from one disastrous energy source to a slightly less disastrous one will of course result in a lowering of emissions.

But perhaps the most dangerous misconception about the

climate crisis is that we have to "lower" our emissions. Because that is far from enough. Our emissions have to stop if we are to stay below 1.5–2 degrees Celsius of warming. The "lowering of emissions" is of course necessary but it is only the beginning of a fast process that must lead to a stop within a couple of decades, or less. And by "stop" I mean net zero – and then quickly on to negative figures. That rules out most of today's politics.

The fact that we are speaking of "lowering" instead of "stopping" emissions is perhaps the greatest force behind the continuing business as usual. The UK's active current support of new exploitation of fossil fuels – for example, the UK shale gas fracking industry, the expansion of its North Sea oil and gas fields, the expansion of airports as well as the planning permission for a brand new coalmine – is beyond absurd.

This ongoing irresponsible behaviour will no doubt be remembered in history as one of the greatest failures of humankind.

People always tell me and the other millions of school strikers that we should be proud of ourselves for what we have accomplished. But the only thing that we need to look at is the emission curve. And I'm sorry, but it's still rising. That curve is the only thing we should look at.

Every time we make a decision we should ask ourselves; how will this decision affect that curve? We should no longer measure our wealth and success in the graph that shows economic growth, but in the curve that shows the emissions of greenhouse gases. We should no longer only ask: "Have we got enough money to go through with this?" but also: "Have we got enough of the carbon budget to spare to go through with this?" That should and must become the centre of our new currency.

Many people say that we don't have any solutions to the climate crisis. And they are right. Because how could we? How do you "solve" the greatest crisis that humanity has ever faced? How do you "solve" a war? How do you "solve" going to the moon for the first time? How do you "solve" inventing new inventions?

The climate crisis is both the easiest and the hardest issue we have ever faced. The easiest because we know what we must do. We must stop the emissions of greenhouse gases. The hardest because our current economics are still totally dependent on burning fossil fuels, and thereby destroying ecosystems in order to create everlasting economic growth.

"So, exactly how do we solve that?" you ask us – the schoolchildren striking for the climate.

And we say: "No one knows for sure. But we have to stop burning fossil fuels and restore nature and many other things that we may not have quite figured out yet."

Then you say: "That's not an answer!"

So we say: "We have to start treating the crisis like a crisis – and act even if we don't have all the solutions."

"That's still not an answer," you say.

Then we start talking about circular economy and rewilding Nature and the need for a just transition. Then you don't understand what we are talking about.

We say that all those solutions needed are not known to anyone and therefore we must unite behind the science and find them together along the way. But you do not listen to that. Because those answers are for solving a crisis that most of you don't even fully understand. Or don't want to understand.

You don't listen to the science because you are only interested in solutions that will enable you to carry on like before. Like now. And those answers don't exist anymore. Because you did not act in time.

Avoiding climate breakdown will require cathedral thinking. We must lay the foundation while we may not know exactly how to build the ceiling.

Sometimes we just simply have to find a way. The moment we

decide to fulfil something, we can do anything. And I'm sure that the moment we start behaving as if we were in an emergency, we can avoid climate and ecological catastrophe. Humans are very adaptable: we can still fix this. But the opportunity to do so will not last for long. We must start today. We have no more excuses.

We children are not sacrificing our education and our childhood for you to tell us what you consider is politically possible in the society that you have created. We have not taken to the streets for you to take selfies with us, and tell us that you really admire what we do.

We children are doing this to wake the adults up. We children are doing this for you to put your differences aside and start acting as you would in a crisis. We children are doing this because we want our hopes and dreams back.

I hope my microphone was on. I hope you could all hear me.'

Further Reading & Useful Websites

Our House is on Fire: Scenes of a Family and a Planet in Crisis by Malena & Beata Emman and Svante & Greta Thunberg (2020)

No One Is Too Small To Make A Difference by Greta Thunberg (2018)

RESURGENCE

at the heart of Earth, Art & Spirit

The Resurgence Trust is an educational charity and the publisher of *Resurgence & Ecologist* magazine. The Trust promotes ecological sustainability, social justice, ethical living and spiritual values, and brings together a community of like-minded individuals who all believe that a more sustainable way of life is possible.

Resurgence & Ecologist, published six times a year, is the UK's only independent publication that focuses on reverential ecology and the arts. *Resurgence* – which merged with *The Ecologist* in 2012 – has been edited by Satish Kumar for almost forty years.

Becoming a member of The Resurgence Trust connects you to the things that really matter: Earth, Art and Spirit. Visit www.resurgence.org for more details.

Resurgence & Ecologist encourages inspiration for a more beautiful world where soil, soul and society are in harmony with each other.

ACKNOWLEDGEMENTS

The seeds of my own activism were sown into the soil of my very soul by my late mother who allowed me to pursue the path I have since followed from the very early years of my life. And so I would wish to express the deepest gratitude to her.

The second woman of great inspiration is my life partner, June. For the past forty years, she has been by my side, like a rock, and without her support and her loving presence, this book would not have been possible.

My heartfelt thanks also go to Susan Clark. She was my wonderful colleague and editor at *Resurgence & Ecologist* magazine and she played an important role in supporting the creation of this book; her eye for detail and skilful communication with the contributors and publisher kept the project on track.

I also want to express my profound appreciation of Emma Randall; my delightful colleague who was ever-ready to help me with the correspondence to our contributors, chasing them to meet their copy dates and somehow managing to keep everybody happy! So thank you Emma for all your hard work and support.

Elaine Green is my tireless and ever-vigilant PA who co-ordinates all my work and helps me remain sane. So again, thank you to Elaine for her patience and perseverance.

This book was only made possible because of Monica Perdoni, the senior commissioning editor at Leaping Hare Press. It was Monica who took my vision for a book on activism and made it manifest in the physical form. She then passed the baton to Katie Bond, publisher at Aurum, to create this new paperback edition. It has been a pleasure to work with them and their teams, especially Jayne Ansell, Peter Bridgewater, Stephen Paul and Phoebe Bath.

Above all, my sincere thanks to all the activist contributors to this book who all found time in their very busy lives to write and share their own stories. I have been privileged and honoured to have these friends in my life and to be part of their activism, their vision, their commitment, their passion and their courage.

The publisher would like to thank the following people and organizations for the use of their portraits:

Satish Kumar: Courtesy of Resurgence; Franny Armstrong: Eamonn McCabe; Bob Brown: AAP Image/Lukas Coch; Helen Beynon: © K Baczkowska; Deepak Chopra: Mark Peterson/ Corbis; Tim Flannery: Eamonn McCabe; Jane Goodall: Jean Kern; Roger Hallam, Polly Higgins: Ian MacKenzie; Caroline Lucas: © David Levene/eyevine; Bill McKibben: © Corey Hendrickson/Polaris/eyevine; Carlo Petrini: © Alberto Peroli/ Courtesy of Slow Food International; Vandana Shiva: Rachelle Hacmac/*Oregon Daily Emerald*; Greta Thunberg: © Anders Hellberg.